# Bovine Maternal Support and Embryo Survival

Aureliano Hernández

# Bovine Maternal Support and Embryo Survival

⁜ Springer

Aureliano Hernández
Facultad de Medicina Veterinaria y de Zootecnia
Universidad Nacional de Colombia
Bogotá, Colombia

ISBN 978-3-031-62390-5        ISBN 978-3-031-62391-2    (eBook)
https://doi.org/10.1007/978-3-031-62391-2

This Springer imprint is published by the registered company Springer Nature Switzerland AG
The registered company address is: Gewerbestrasse 11, 6330 Cham, Switzerland

If disposing of this product, please recycle the paper.

# Preface

The present book is about maternal preparation to nourish the bovine embryo, how nature provides its survival in the uterus and relevant causes of embryonic mortality.

It covers embryological, morphological and physiological relevant events in a general manner, with the aim of giving the interested reader some facts which motivate the study of basic reproductive events and possibly research pathways as well as some practical applications. Some genetic relevant aspects are presented in a view of giving a somewhat ample picture.

The present general approach is mainly intended to be used by veterinary students and professionals. Graduate students and researchers might find in the book some useful issues.

Bogotá, Colombia                                               Aureliano Hernández
October 2023

# Contents

# Chapter 1
# Embryology of the Reproductive Tract

**Abstract** The intrauterine development of the mammalian female genital organs is presented, with special reference to the bovine, within the frame of morphological, physiological, and genetical aspects. The chronological discussion of early embryological events leads to the exposure of specific mesodermal, mesonephric and paramesonephric changes, gonadal formation and primordial cells migration, and subsequent ovarian and female tract developmental changes. Most processes are illustrated by pictures and drawings.

**Keywords** Embryology · Bovine · Genes · Mesonephros · Paramesonephric ducts · Primordial germ cells · Ovary · Oviduct · Uterus · Cervix · Vagina · Genes · Molecules

## Early Embryonic Events

After fertilization and the appearance of the two first new cells (blastomeres Figs. 1.1 and 1.2), proliferation takes place and the cell number increases as to become an embryo with 4, 8, 16, 32, 64, 128, .... These group of cells are known to be at the morula stage, given their external look.

The first mitotic cell divisions occur during the embryos migration through the oviduct and the correspondent entry in the uterus. The event latter happens at days 4 or 5 post-insemination. The embryo is still surrounded by the zona pellucida (ZP; Figs. 1.2 and 1.3).

Some of the cells separate and, therefore, a space known as the blastocele appears at the center of the embryo. Whereas a group of cells remain together situated at a pole and become the internal cell mass (ICM). Towards the 8 days of gestation, the ZP is thinned and finally ruptures. The most external layer of the ICM is called the Rauber's layer, which covers the epiblast but at the top of the epiblast. The most internal line of cells in the ICM is known as the hypoblast. Later on, the Rauber's layer disintegrates and the hypoblast is then part of the ectoderm. The endoderm derived from the ICM is placed in the internal part of the ZP as a single layer of

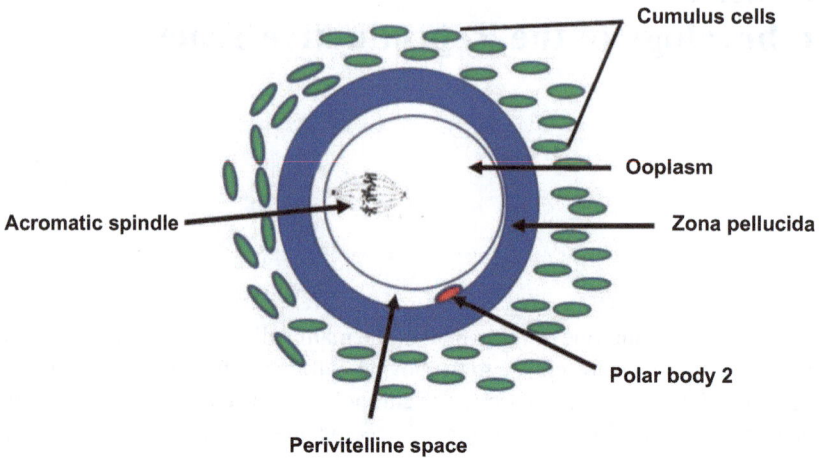

**Fig. 1.1** An ovum surrounded by the zona pellucida and some cumulus cells

**Fig. 1.2** Two blastomeres and zona pellucida. It corresponds to a 2-day-old bovine embryo

**Fig. 1.3** Morula. Arrow: zona pellucida. It corresponds to a 4 day-old-bovine embryo

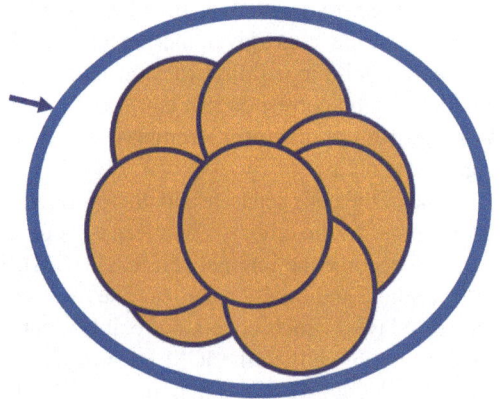

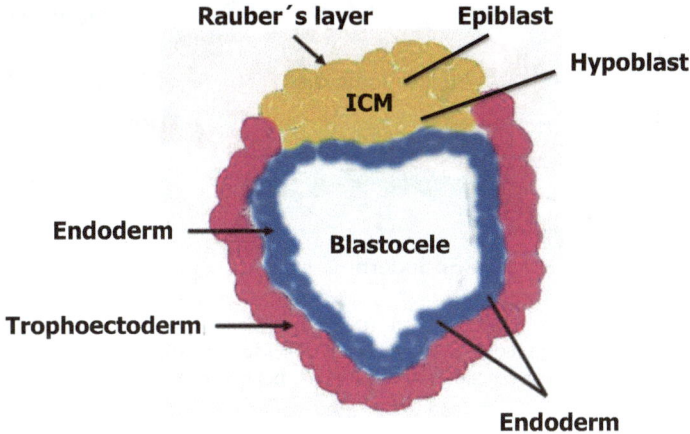

**Fig. 1.4** Scheme of a 9-day-old bovine blastocyst. ICM: internal cell mass. Note that the zona pellucida has been discarded

cells. The space englobed by the endoderm is the blastocele (Fig. 1.4). The embryo is known then as the blastocyst, which becomes spherical, then ovoid, and filamentous afterwards.

At the end of the morula stage, a group of cells separate and migrate to one pole of the embryo. That group is called the internal cell mass (ICM). The remaining cells form a layer located in the internal zone of the zona pellucida, known as trophoblast (TR), ectoderm or tropho-ectoderm. The upper zone of the ICM is known as the epiblast and the internal cells constitute the hypoblast (Fig. 1.4).

The dorso-ventral body axis is established when the ICM shows up.

The epiblast gives rise at the future posterior zone of the embryo's body, to a cell cumulus, the primitive streak (PS Fig. 1.5). The primary endodermic and mesodermal cells come from the PS and the germinative (germinal) cells. The outcome of the PS indicates the establishment of the antero-posterior body axis. The growth and expansion of the ICM result in the origin of the embryonic disk, which will be the embryos body (Fig. 1.5).

The mesoderm expands in multiple directions. His position becomes between the ecto- and endoderm. It becomes a thick structure located at both sides of the longitudinal body axis. There, it is named the dorsal (paraxial) mesoderm, which becomes thinner as the intermediate mesoderm. The latter divides in two layers which together form the lateral mesoderm. The external layer (somatic mesoderm) accompanies the trophoectoderm and together are known as the chorion. The internal one (the splanchnic or visceral mesoderm) covers the endoderm; this two layered structure is known as the splanchnopleure (Fig. 1.6).

The dorsal mesoderm will give rise to the somite, which will arise in pairs in the anterior part of the embryo's body and in sequence towards the caudal zone. From them derive the vertebrae, the striated muscles, and the dermis. From the intermediate mesoderm, the major part of the urogenital system develops.

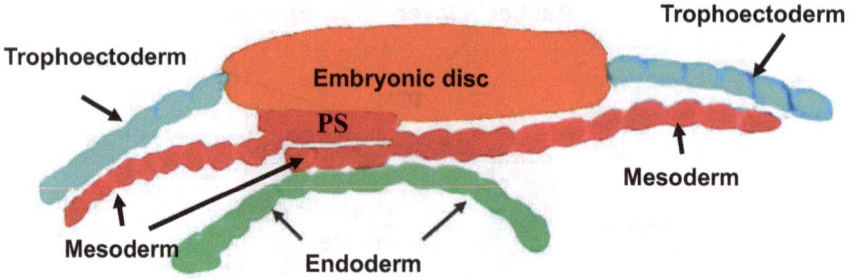

**Fig. 1.5** The primitive streak (PS) is shown at the future caudal zone of the embryonic disk. The mesodermic cells arise from the PS and will occupy zones located between the ecotoderm and the endoderm. This scheme corresponds to events which belong to the third week of gestation in the bovine

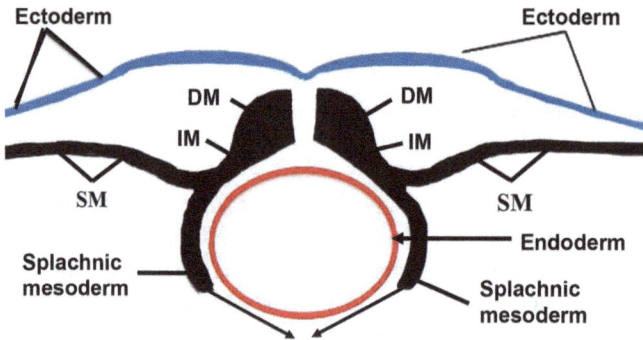

**Fig. 1.6** The dorsal mesoderm (DM; also named paraxial mesoderm) located at both sides of the embryo's longitudinal axis is thick. At each side, it becomes the intermediate mesoderm (IM), which continues as the lateral mesoderm which branches in two parts: the external somatic mesoderm (SM) and the internal one known as the splanchnic mesoderm. The SM accompanies the ectoderm. Together will be called chorion (somatopleure). Eighteenth day of gestation

An extensive number of genes have been discovered as a key to the establishment of the genitalia. The sexual dimorphism is in the male's Y chromosome, which, among other genes, possesses the SRY gene. Its protein product is a transcription factor for the expression of determinant several genes for the differentiation of the genitalia. Hence, the SRY protein leads the development of the male genitalia, and its absence directs the correspondent female organs (Sadler 2019).

## The Excretory System

Initially, in mammals, the intermediate mesoderm is converted in the pronephros, which briefly functions in some species, and is replaced by another transitory system—permanent in avian species—the mesonephros. Later on, the definitive excretory system, the metanephros derives from the mesonephric duct (Figs. 1.7, 1.8, 1.9, and 1.10).

The intermediate mesoderm is specified due to a decreasing gradient of the bone morphogenetic protein (BMP) initiated in the lateral mesoderm and from unknown cell signals from the dorsal mesoderm. The response to the abovementioned processes is the expression of transcription factors Pax-2 and Pax-8. Both of them induce LIM-1 (LHX-1) in the intermediate mesoderm. LIM-1 is needed for the aggregation of cells destined to form primary nephric tubules known as pronephros (Carlson 2018). The pronephros is composed by tubules which filtrate blood and flow the resulting liquid into the pronephric duct. The same filtration apparatus functions in both the mesonephros and metanephros (Fig. 1.7).

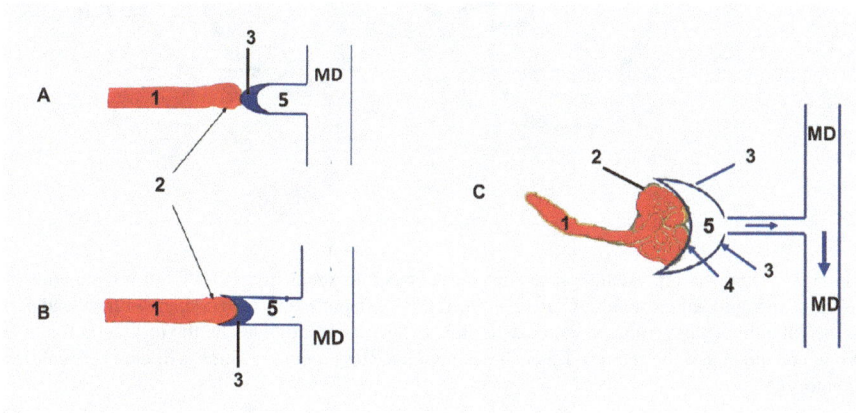

**Fig. 1.7** (**a**) At first, intermingled capillaries (2) derived from an arteriole (1), invaginate the blind end a mesonephric tubule (3). (**b**) The external wall (4) of the latter covers the capillaries. (**c**) Liquid excretions from the intermingled capillaries will pass into the light of the mesonephric tubule and run into the mesonephric duct (MD). The external wall of the mesonephric tubule becomes the visceral part of the renal corpuscle and the parietal layer will be the non-invaginated wall of the mesonephric tubule

**Fig. 1.8** Within the mesonephros, intermingled capillaries (1) are in contact with mesonephric tubules (2) as shown in in the oval (3)

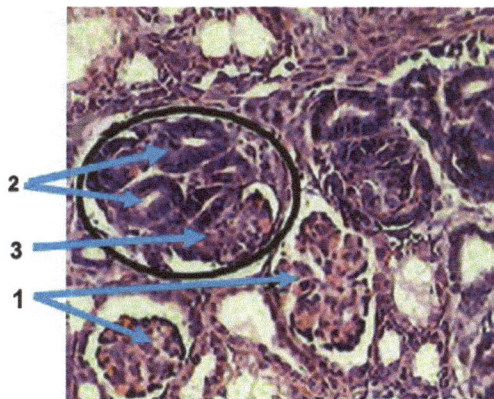

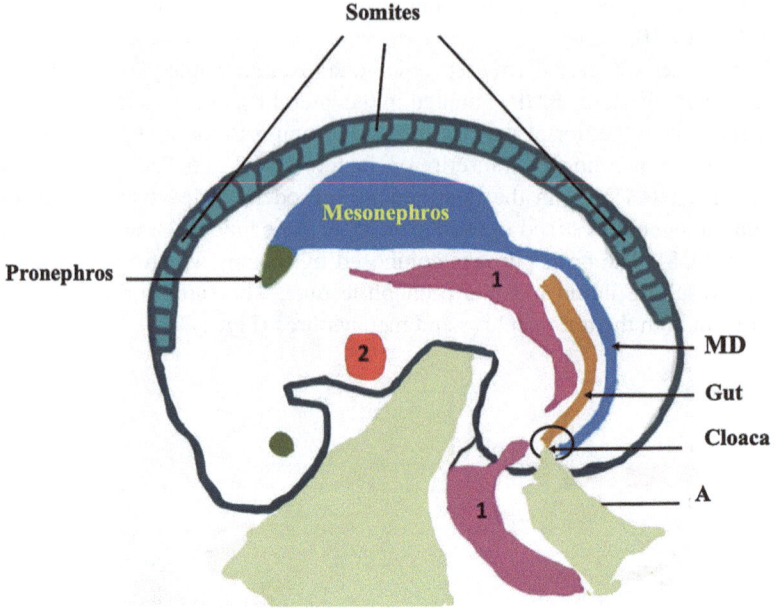

**Fig. 1.9** Schematic representation of some structures and zones in a 19-day-old bovine embryo 19 days. 1: Angiogenic zones. 2: Cardiac atria. The mesonephric duct is in contact with the end part of the gut, which will be named the cloaca. This, in turn, will be divided by the uro-rectal fold in a urogenital sinus and the rectum. Later in development the urogenital sinus will connect with the allantois (A)

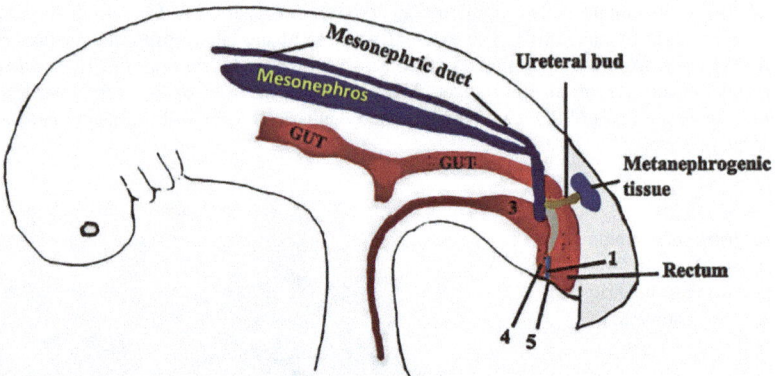

**Fig. 1.10** The uro-rectal fold (1) divides of cloaca (2) in the rectum and the urogenital sinus. The ureteral bud emerges from the mesonephric duct. The metanephrogenic tissue, originated from the mesenchyme, gives rise to the metanephric kidney

The filtration unit is composed by blood capillaries which result from branching of an arteriole. The capillaries are then intermingled and invaginate the external wall of a tubule, which communicates with the duct (Fig. 1.7). Thus, as the invagination progresses, the external wall of the tubule covers the intermingled capillaries (glomerulus) and a space is left between the external and the internal tubule walls. The internal tubule wall will become the parietal layer of the filtration unit (renal corpuscle) (Figs. 1.7 and 1.8).

The mesonephric ducts opens into the cloaca, the end part of the primitive gut. The cloaca has cloacal membrane at its end caudal zone (Figs. 1.9 and 1.10).

The mesonephros functions during a restricted period and is replaced by the definitive kidney, the metanephros. The ureteral bud, which comes from the mesonephric duct, will be covered by the metanephrogenic tissue and, together, will constitute the urinary system (Fig. 1.10).

## The Gonads

At first, the genital ridge presages the appearance of the gonad at the ventral-caudal zone of the mesonephros, as a proliferation of the coelomic epithelium. The primordial germinative cells (GC germinal cells), the oogonia predecessors in the female (with XX sex chromosomes) and the spermatogonia in males (with XY chromosomes), appear in the epiblast from the extra embryonic morphogenetic protein 4 (BMP4). The GC migrate through the primitive streak to the allantoic mesoderm and then to the endoderm of the yolk sac. They will migrate to the gonad, along the body wall. During migration, they exhibit a high mitotic rate induced by the factor LIF (leukemia inhibitory factor). The steel factor, a ligand for the kinase protein c-Kit, also contributes to the abovementioned proliferation. Las CG appear in the bovine gonad at the 25th day of gestation (Figs. 1.11 and 1.12).

The epithelium of the posterior and ventral part of the mesonephros gives rise to the genital ridge, the future gonad. The GC are needed to enter the gonad, as to induce its formation (Sadler 2019). They are histochemically recognized for their alkaline phosphatase-positive expression and the pluripotential transcription factors such as OCT4, which is responsible for the GC activity. One to two thousand GC enter the genital crest. The gonad is established by the expression of WT-1 in the intermediate mesoderm and that of LIM-1 and steroidal gene factor 1 (Sinowatz 2010).

The gonad will be transformed to an ovary by the expression of DAX-1, in the absence of SRY (Gene in the Y chromosome; Sinowatz 2010).

The GC concentrate in the cortical gonadal zone (in the horse, this happens in the medulla). The medulla also contains GC grouped as primitive sex chords, although their development is lower than in the male. In the early stage of gonadal development, the gonadal epithelium proliferates and penetrates to establish the primitive sex chords, which are in an indifferent sexual stage yet.

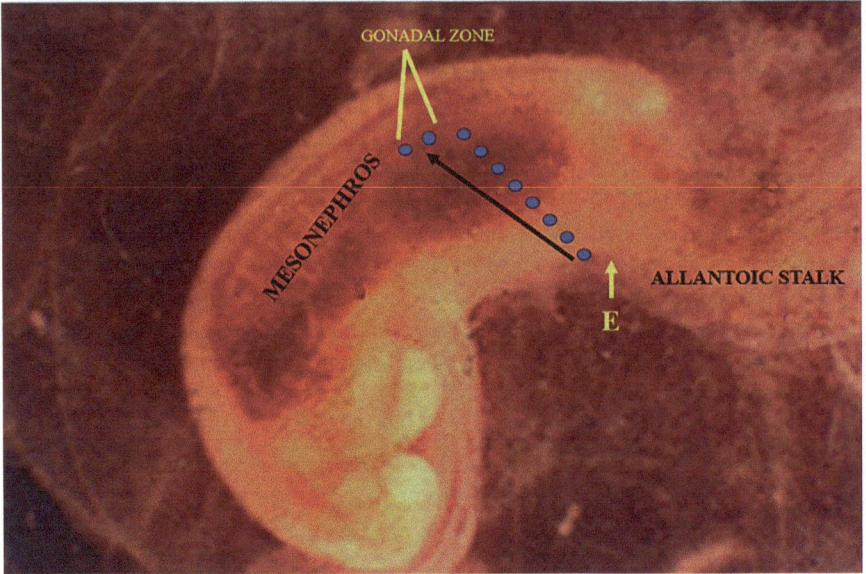

**Fig. 1.11** The primordial germ cells migrate through the mesentery from the endoderm of the midgut, to the genital ridge, the gonad antecessor. Bovine embryo, 25 days

**Fig. 1.12** The bovine gonad (arrow) appears attached to the mesonephros (oval)

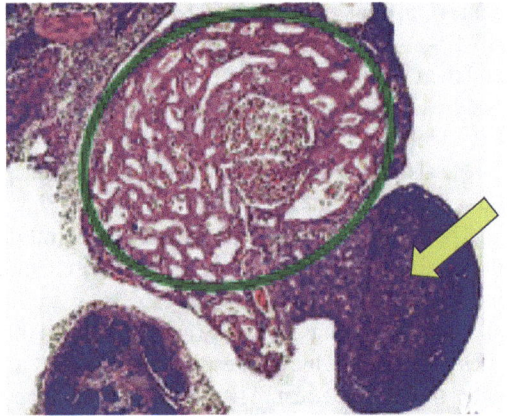

The more developed gonad then contains: (1) mesenchyme, (2) mesonephric tubules, (3) superficial epithelium, and (4) GC. In more advanced stages, more blood vessels, nervous elements, and lymphatics should be established therein.

The GC are surrounded by the sex chord cells, derived from the mesonephric tubules and/or the celomic (superficial) epithelium. This new structure is known as the primordial follicle.

At around the 35th day of gestation, the gonad is bipotential. The SRY expression in the male embryo reaches maximal expression, and at day 43 the gonad is at

an early stage of differentiation. Between days 35 and 43 there are important differences in the types of expressed genes, according to the correspondent sex; 767 differential genes in males and 545 in females between days 35 and 39, and 3157 in
males and 2008 in females in the range of 39 and 43 days of pregnancy (Planells
et al. 2020). Likewise, sexual metabolic differences were encountered, which could
result in the inactivation of the X chromosome in male embryos, process that takes
place in only one of the X females chromosomes. Consequently, there would be
more expression of genes and transcripts in the female, related to sex-dependent
different metabolic substrates, specially, those involved in glycolysis and the pen-
tose phosphate cycle (Bermejo-Alvarez et al. 2011).

An unexpected outcome is the concourse of gene IGF2 in the formation of ovarian follicle cells in mice (Keshet et al. 2023).

The GC form nests and are transformed as oogonia. Some oogonia are encapsu-
lated in sex chords, others are lost, and many of them enter the prophase of the first
meiotic division and get arrested in the pachytene stage. When arrest fails in some
of these cells, the remaining ones make up an ovarian reserve (Strauss III and
Williams 2019). These oogonia are surrounded by follicle cells to become the primordial follicles (Fig. 1.13).

Many of the somatic cells in the XX gonad are to be converted in the ovarian
follicular and the theca cells. The follicular (granulosa) cells are differentiated in
two non- synchronized ways. The first population, which will become a medullar
group, are specified by upregulation of the cell cycle inhibitor Cdkn1b/p27 and the
subsequent induction by FOXl2. The second cell population express the RSPO1
Lgr5 receptor to augment the cell reserve, which occur before the concourse of
Cdkn1b/p27 and Foxl2 to induce cell cycle arrest. The abovementioned cells give
rise to medullar and cortical follicular cells. The progenitor cells in the stroma stay
as undifferentiated Lhx+ cells and express Arx, CouptfII/Nr2f2 and Mafb. At the
end of gestation, the Hedgehog signals coming from follicle cells convert the progenitor cells in steroidogenic thecal cells Gli+. Some stromal cells of mesonephric
origin become theca cells Gli+. Some stromal cells remain in the ovary after birth
as non-steroidogenic which express CouptfII and Arx, as do the Leydig cells in the
testis (Rotgers et al. 2018).

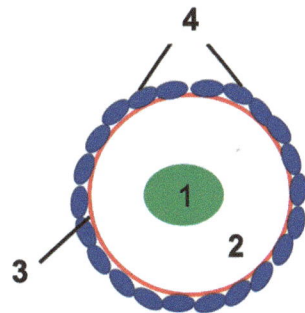

**Fig. 1.13** Primordial
follicle. Primary oocyte
surrounded by flat
follicular cells (4). 1.
Nucleus. 2. Ooplasm. 3.
Vitelline membrane

## Prenatal Development of the Genital Tract

The paramesonephric (Müller) ducts appear as longitudinal invaginations of the celomic epithelium (Fig. 1.14) along the mesonephric folds, laterally to the mesonephric ducts.

The paramesonephric caudally ducts fuse to form the oviduct, the uterus, the cervix, and the anterior portion of the vagina. In the different domestic mammals, the ducts fusion occurs at different levels and obeys to diverse mechanisms (Major et al. 2022).

Uteri are classified as simplex, like non-humans, bicornual in mares, ruminants, and domestic carnivora (Figs. 1.15, 1.16, and 1.17).

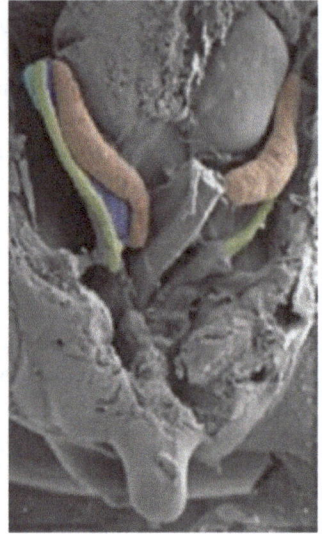

**Fig. 1.14** Scanning electron photograph of the ventral view of an embryo, to show the gonad (G), mesonephric duct (M), and paramesonephric (Müller) duct (P). (With kind permission by Dr Kathleen Sulik)

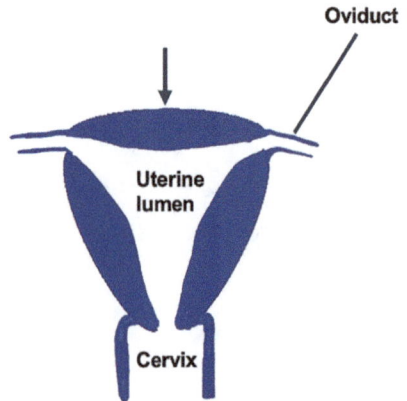

**Fig. 1.15** Schematic drawing of a simplex uterus. The arrow shows the point of fusion of the paramesonephric ducts

**Fig. 1.16** Bicornual
uterus. Mare. 1: ovary.
2: uterine horns. 3: cervix.
4: vagina

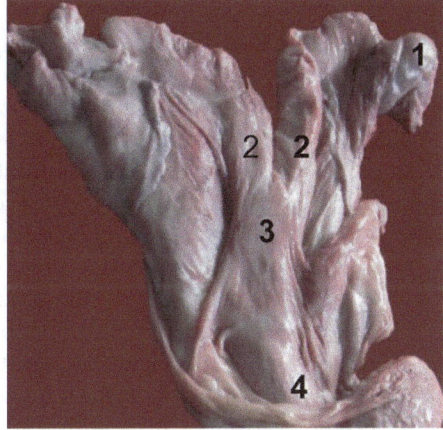

**Fig. 1.17** Pregnant
bicornual uterus, Sheep.
The line shows the plane of
fusion of the
paramesonephric ducts.
Note the outgrowths (most
of them with melanin),
where the embryo/fetus are
known as caruncles. 1:
uterine horns. 2: uterine
body. 3: part of the cervix

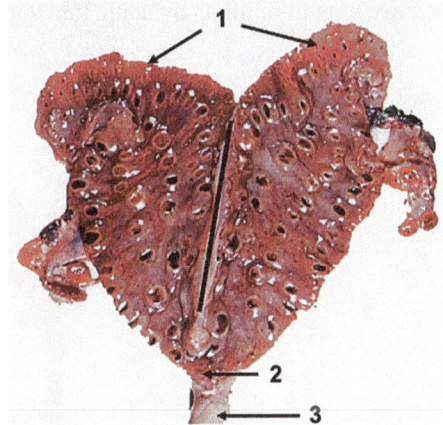

Placodes near the mesonephric duct express Lhx-1 and acquire a chord-like appearance and extend towards the mesonephric ducts under the influence of Wnt-4 produced in the mesonephros. The extremities of the paramesonephric ducts make up a proliferative center and due to the intervention of Wnt-9b from the mesonephric ducts; they progress caudally towards the urogenital sinus. When they contact the latter structure, they develop a lumen. The cranial extremity of the paramesonephric duct opens into the celomic cavity as a funnel like end (Carlson 2018).

The commencement of paramesonephric duct formation needs Wnt7a signals from the ducts epithelium. Those signals are linked to the expression of Hox genes (Hoxd-10 a Hoxd-13 and Hoxa paralogues), which are extended in the human genital apparatus.

Hoxa-9 is in the oviduct; Hoxa-10 in the uterus expressed Hoxa-11 in the utero and cervix and Hoxa-12 in the cranial part superior of the vagina (Carlson 2018).

It is to be noted that genes expression during the development of genitalia in humans are extremely similar to the same process in the cow (Ross et al. 2009).

In the genital tract, the well-known inductive mechanism exerted by the mesenchyme on the epithelium as to define the type of tissue is evident in the genital tract. Thus, if the vaginal adjacent mesenchyme is placed in contact with the uterine primordium, the uterine epithelium will be stratified as in the vagina and not the simple cylindrical or pseudo-stratified normally seen in the uterine lining (Cunha 1976).

In males, the SrY protein codifies for the production of the anti-Müllerian hormone (AMH or anti-paramesonephric). The AMH absence in the female permits the development of the paramesonephric ducts (Fig. 1.18).

Regression of mesonephric ducts in the female occurs by the action of COUP-TF II (chicken ovalbumin upstream promoter transcription factor 2), which is expressed in the mesenchyme as to inhibit the fibroblast growth factor. Hence, in the female, only mesonephric duct rudimentary structures are left. Hox genes mutations produced homeotic transformations. In the absence of Hoxa-10, the cranial uterine zone becomes an oviduct. In contrast with other body organs, the Hox genes persist throughout life span. This might be related to the necessary plasticity of the reproductive tract to fulfill its dynamic functions.

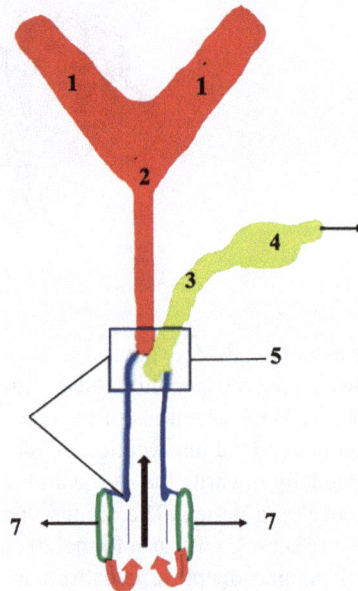

**Fig. 1.18** Morphogenesis of the vagina. Schematic representation. The paramesonephric ducts fuse to become the bicornual uterus (1), the cervix (2), and the anterior vagina. Its central zone is formed by the urogenital sinus (3) derived sinovaginalis bulb, fused with the end part of the paramesonephric ducts (frame). The vaginal vestibule or caudal zone of the vagina is the result of the invagination carried out by the genital pouches (7) as to lined up with the middle vaginal channel (arrows). The urogenital sinus content will drain into the future urinary bladder and this into the allantoic cavity (small arrow, right)

The paramesonephric ducts then, as already mentioned, fuse to give rise to the uterus, cervix, and part of the vagina. The broad ligament is the result of the displacement of the uterus, which conveys the genital crest tissue (Carlson 2018).

## External Genitalia

At first, the genitalia in males and females are anatomically similar (Fig. 1.19).

The cloacal membrane, which separates the cloaca from the exterior, should disappear. The urethral folds will be carried into the vaginal channel to form the labia minora and the genital pouches will be the labia majora of the vulva approximate to each other. The genital tubercle does not grow as in the males and will occupy an internal place (Figs. 1.19, 1.20, and 1.21). In the end, the only visible structures from the outside will be the labia majora of the vulva.

The genital tubercle develops from the mesenchyme. During the indifferent stage, hormones do not influence the development of the genital tubercle, but later on androgens and estrogens influence its conversion as the penis.

In humans, the genital tubercles in both sexes have cells which have androgens and express 5α-reductase, but only in males, the tubercle elongates, in virtue of the greater amount of testosterone produced by the interstitial cells in the gonads of male embryos.

The genital tubercle express the 5′ elements together with the gen Hox family, mainly Hoxa-13 and Hoxd-13. However, the signal needed to initiate the genital tubercle development is not known as yet.

The Shh expressed in the endodermic epithelial plaque of the urethra is the main necessary molecule for the genital tubercle growth, through its action on the mesenchyme and ectoderm. There are many FGF signaling molecules and Wnt in the genital tubercle, but their functions remain unknown (Carlson 2018).

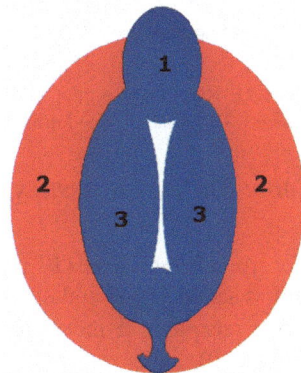

**Fig. 1.19** Schematic representation of the external genitalia. Indifferent stage. 1: Genital tubercle. 2: Urethral folds. 3: Genital pouches. Cloacal membrane: the space between the urethral folds

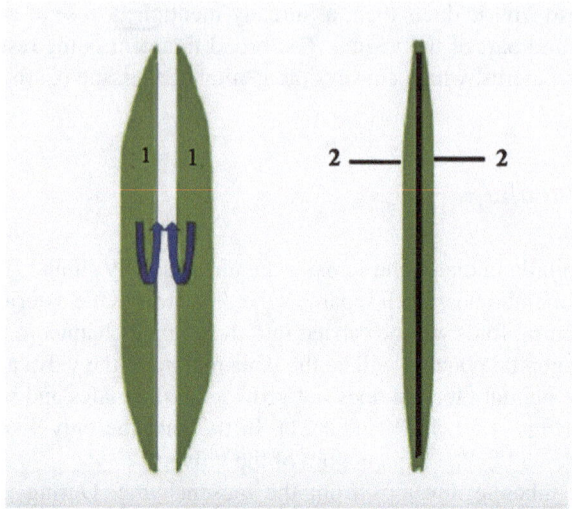

**Fig. 1.20** Drawing (left) to explain the change undergone by the genital pouches as to become the vulva. The arrows indicate the invagination of the tissues separating the genital pouches. The drawing at right corresponds to the final approach of the pouches which then become the external vulvar folds. Between them the channel is depicted as a darker line

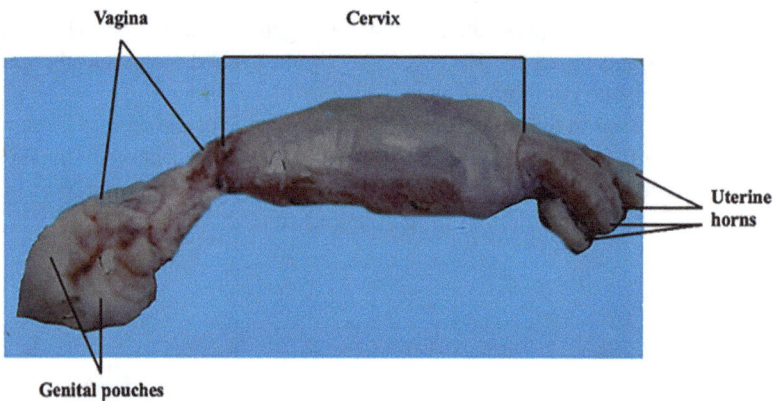

**Fig. 1.21** Reproductive tract. Bovine fetus. Note that the cervix is distended (due to its aqueous content). The small uterine horns are rolled up

## *Some Endocrine Issues*

The aromatase enzyme (17-β dehydrogenase), which converts androgens 17 β estradiol, is active in the fetal ovary during the third month of gestation. There are estrogen α receptors in the ovary to facilitate the corresponding response (Burkhart et al. 2010).

Some brain nuclei can be changed by sexual hormones impact. This explains the so-called sexual dimorphism. Gorski et al. (1978) found, in fetal rats, that the size of the medial preoptic hypothalamic nucleus was 8 times bigger than that of the females. When the adult males were castrated, the nucleus decreased in size.

Estradiol implants in the medial preoptic area and the addition of dihydrotestosterone activated male behavior in male rats (Davis and Barfield 1979; Bakker and Baum 2008). Possibly, by expression and of the gonadotrophin releasing hormone (GnRH), which is active during the fetal life (Ebling 2005).

In female embryos, this phenomenon could be absent. If so, this might be one mechanism of sex determination, so-called "brain sex" (Mcewen 1980; Short 1972).

While androgens are produced in the fetal testis during, estrogens are not in the female, since complete follicle development only occurs at puberty.

In the female, the mesonephric duct atrophies given the insufficient androgens elaboration.

The biological olfactory stimuli act in reproduction in mammals, attributed to pheromones synthesis in sweat glands, among them androstenedione (Verhaeghe et al. 2013). During fetal development, GnRH neurons migrate from the diencephalon to the olfactory region in fetal rats, and maintain their morphological connections with the hypothalamus. These cells would be responsible for capturing pheromones (Wray et al. 1989).

# References

Bakker J, Baum MJ (2008) Role for estradiol in female-typical brain and behavioral sexual differentiation. Front Neuroendocrinol 29(1):1–16

Bermejo-Alvarez P, Rizos D, Lonergan P et al (2011) Transcriptional sexual dimorphism during preimplantation embryo development and its consequences for developmental competence and adult health and disease. Reproduction 141:563–557

Burkhart MN, Juengel JL, Smith PR et al (2010) Morphological development and characterization of aromatase and estrogen receptors alpha and beta in fetal ovaries of cattle from days 110 to 250. Anim Reprod Sci 117(1–2):43–54

Carlson BM (2018) Human embryology and developmental biology, 6th edn. Elsevier, New York. Spanish edition: Carlson BM (2020) Embriología humana y biología del desarrollo, 6ta edn. (trans: Peña AM, Viejo F). Elsevier, Barcelona

Cunha GR (1976) Stromal induction and specification of morphogenesis and cytodifferentiation of the epithelia of the Mullerian ducts and urogenital sinus during development of the uterus and vagina in mice. J Exp Zool 196:361–370

Davis PC, Barfield RJ (1979) Activation of masculine sexual behaviour by intracranial estradiol benzoate implants in male rats. Neuroendocrinology 28:217–227

Ebling FJ (2005) The neuroendocrine timing of puberty. Reproduction 129:675–683

Gorski RA, Gordon JH, Shryne JE et al (1978) Evidence for a morphological sex difference within the medial preoptic area of the rat brain. Brain Res 148:333–346

Keshet G, Bar S, Sarel-Galili L et al (2023) Differentiation of uniparental human embryonic stem cells into granulosa cells reveals a paternal contribution to gonadal development. Stem Cell Reports 18(4):817–828. https://doi.org/10.1016/j.stemcr.2023.03.004

Major AT, Esterman MA, Roly SY et al (2022) An evo-devo perspective of the female reproductive tract 16. Biol Reprod 106(1):9–23

Mcewen BS (1980) Gonadal steroids and brain development. Biol Reprod 22:43–48

Planells B, Gómez-Redondo I, Sánchez JM et al (2020) Gene expression profiles of bovine
    genital ridges during sex determination and early differentiation of the gonads. Biol Reprod
    102(1):38–52

Ross DGF, Bowles J, Hope M et al (2009) Profiles of gonadal gene expression in the developing
    bovine embryo. Sex Dev 3(5):273–283

Rotgers E, Jorgensen A, Hung-Chan Yao H (2018) At the crossroads of fate-somatic cell lineage
    specification in the fetal gonad. Endocr Rev 39(5):739–759

Sadler TW (2019) Langman's medical embriology, 14th edn. Wolters Kluwer, Philadelphia.
    Spanish edition: Sadler TW (2019) Embriología médica de Langman, 14ta edn. Wolters
    Kluwer Health, Bogotá

Short RV (1972) Sex determination and differentiation. In: Austin CR, Short RV (eds) Reproduction
    in mammals: 2 Embryonic and fetal development, 2nd edn. Cambridge University Press,
    Cambridge, p 93

Sinowatz F (2010) Development of the urogenital system. In: Hyttel P, Sinowatz F, Veljsted M
    (eds) Essentials of domestic animal embryology, 1st edn. Elsevier, Edinburgh, p 268

Strauss JF III, Williams CJ (2019) Ovarian life cycle. In: Strauss JF III, Barbieri RL, Gargiulo AR
    (eds) Yen & Jaffes reproductive endocrinology, 8th edn. Elsevier, Amsterdam, p 495

Verhaeghe J, Gheysen R, Enzlin O (2013) Pheromones and their effect on women's mood and
    sexuality. Views Vis Obgyn 5(3):189–95

Wray S, Grant P, Gainer H (1989) Evidence that cells expressing luteinizing hormone-releasing
    hormone mRNA in the mouse are derived from progenitor cells in the olfactory placode. Proc
    Natl Acad Sci USA 86:8132–8136

# Chapter 2
# Neuroendocrine Control of Ovarian Function

**Abstract** In this chapter, a general approach of the hypothalamic-hypophyseal physiology is considered, in a view of explaining ovarian endocrine function. The origin of the hypophyseal components is presented. Various hypothalamic and hypophyseal functions are outlined, as well as some molecular influences on the hypothalamic functions.

**Keywords** Neuroendocrine cells · Hypothalamus · Hypophysis · GnRH · LH · FSH · Rathke's pouch

Reproductive function depends on influences exerted on the hypothalamus by external variables (climatic and others), as well as the feedback input of molecules generated in target organs indirectly controlled by the hypothalamus (through the hypophysis), including the ovaries.

The information received by neurons in the hypothalamus nuclei, provokes elaboration of releasing factors (hormones) which travel along venous blood to the adenohypophysis. One of them, the gonadotrophin-releasing hormone (GnRH) binds receptors in the basophil cells of the hypophysis, which secrete luteinizing (LH) and follicle stimulant (FSH) hormones. These molecules are the more important substances to control the ovarian function as related to ovulation, fertilization, and pregnancy maintenance.

The hypothalamus is the area composed by nuclei (mainly by neuronal bodies) located behind the optic chiasma, at the base of the third ventricle and its lateral walls. The nuclei are named according to their location. The hypothalamus is connected with many encephalic areas and controls varied biological events, such as appetite, temperature, body weight, growth, some metabolic processes, hydric balance, sleep, emotions, and circadian rhythms (McCartney and Marshall 2019).

The GnrH-secreting neurons are in the medial preoptic area and the arcuate nucleus. They respond to stimuli coming from extrahypothalamic zones and secrete thyrotrophin (TSH), adrenocorticotropin (ADH), FSH, and LH (Fig. 2.1).

The neurons in the supraoptic and paraventricular nuclei produce oxytocin and vasopressin. Their axons terminate in the neurohypophysis, where oxytocin and

**STIMULI: (TEMPERATURE, LIGHT, STRESS ...)**

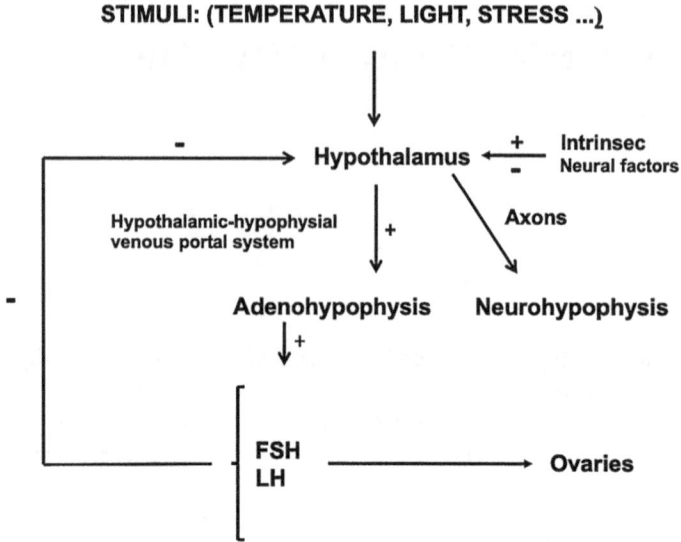

Fig. 2.1 External stimuli are carried to the hypothalamus, which responds by secreting GnRH as to stimulate FSH and LH secretion in the anterior hypophyseal lobe, as well as another hormones. When serum levels of FSH and LH are in adequate levels, these hormones control GnRH output (lines with a – on top). FSH and LH have receptors in thecal and follicular cells in the ovaries, where progesterone and estrogens are synthesized during the ovarian cycle

vasopressin can be identified. Oxytocin is a key molecule in milk ejection, uterine motility during parturition and pregnancy recognition. Vasopressin, as its name implies, participates in controlling systemic blood pressure.

The anterior lobe of the hypophysis (adenohypophysis) contains basophilic and acidophilic cells. The former secrete TSH, ADH, LH, and FSH; the latter, growth hormone and prolactin. The adenohypophysis derives from the embryonic buccal epithelium, which develops a dorsal outgrowth known as the Rathke's pouch. This formation reaches the infundibulum at its anterior area and surrounds it. The non-surrounding part is known as the pars distalis, and that around the infundibulum as the pars tuberalis (Fig. 2.2).

In the preoptic area and median eminence, a beta endorphin (BE, an opioid) produced in neurons, associated with GNRH cells; BE inhibits GnRH production (Leishin et al. 1992).

Hypothalamic neurons, from the preoptic and periventricular areas, stimulate GnRH neurons to promote LH secretion (Jansen et al. 1997; Le et al. 1999).

It was found that noradrenaline has receptors in GnRH neurons, although the biological significance is not elucidated yet (Hosny and Jennes 1998).

In Chap. 4 of this book, some other mechanisms of hypothalamic physiology are presented.

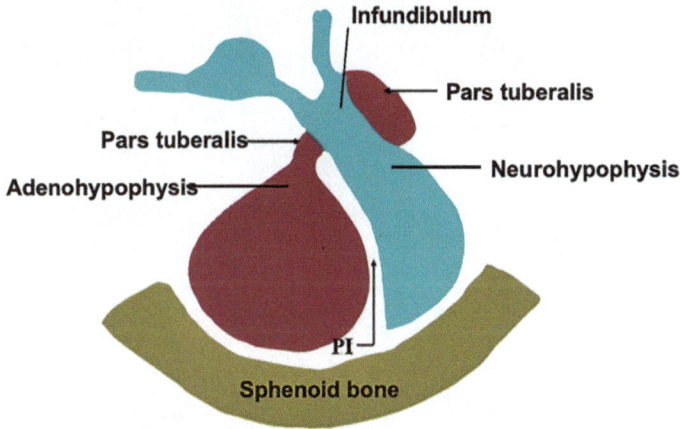

**Fig. 2.2** Schematic representation of the hypophysis. The adenohypophysis has two known lobes: the anterior one (adenohypophysis) and the posterior one, the neurohypophysis. Note that the anterior lobe surrounds the upper part of the infundibulum (pars tuberalis), which gives rise to the posterior lobe. Between the two lobes, there is an intermediate zone (PI)

# References

Hosny S, Jennes L (1998) Identification of alpha(1B) adrenergic receptor protein in gonadotropin releasing hormone neurones of the female rat. J Neuroendocrinol 10(9):687–692

Jansen HT, Hileman SM, Lubbers LS et al (1997) Identification and distribution of neuroendocrine gonadotropin-releasing hormone neurons in the ewe. Biol Reprod 56(3):655–662

Le WW, Berghorn KA, Rassnick S et al (1999) Periventricular preoptic area neurons coactivated with luteinizing hormone (LH)-releasing hormone (LHRH) neurons at the time of the LH surge are LHRH afferents. Endocrinology 140:510–519

Leishin S, Run L, Kraeling R et al (1992) Distribution of beta-endorphin immunoreactivity in the arcuate nucleus and median eminence of postpartum anestrus and luteal phase cows. Neuroendocrinology 56:436–444

McCartney CR, Marshall JC (2019) Central control of reproduction. In: Strauss JF, Barbieri RL (eds) Yen & Jaffe's reproductive endocrinology, 8th edn. Elsevier Health Sciences, Philadelphia, p 36

# Chapter 3
# Follicle Dynamics in the Ovary Before Puberty

**Abstract** Before puberty, including part of the intrauterine life, ovulation mechanisms are not yet ready. They appear during puberty. Morphological aspects of follicle development are presented. Pertinent mechanisms of initial oocyte maturation are discussed in harmony with histological and endocrine events. Meiotic episodes are considered. There are a good number of figures in this chapter.

**Keywords** Oocyte · Meiosis · Ovarian follicles · Transforming growth factor β1 · Insulin growth factor · Anti-Müllerian hormone · 17β estradiol · LH · ADN

Before puberty, including part of the intrauterine life, ovulation mechanisms are not yet ready. They appear during puberty.

At puberty there is a period of irregularity when ovulation could occur, but the endocrinological processes are not stable yet. In fact, before puberty, FSH and LH blood levels are scarce as compared to those at puberty and later on during reproductive life.

In mammals, a biological "waste" of follicles occur (see Chap. 1) and, for instance, in humans, five or six millions of primordial follicles found in a fetal ovary (Fig. 3.1) between half a million and a million survive at birth. Only 400 have the potential of being ovulated during reproductive life (Ford et al. 2020). The control of development of primordial follicles has been attributed to AMH action, although the mechanism is unknown (Yang et al. 2017).

The transition from primordial to primary follicles, in mice, could depend on the concourse of molecules such as the newborn ovary homeobox (NOBOX), spermatogenesis and oogenesis helix-loop-helix 1 (SOHLH1), and SOHLH2, since the ovaries of newborn female mice Nobox_/_ o Sohlh1_/_ contain a similar number of GC nests and primordial follicles than controls, but their progression to secondary follicles is affected (Edson et al. 2009).

The transforming growth factor β and FSH are important in the development of follicular cells and hence for follicular growth (Ingman and Robertson 2009; Li et al. 2022).

**Fig. 3.1** Primordial
follicles. One layer of
flattened cells surrounds
each oocyte. 1: nuclei, 2:
cytoplasm of the oocyte I

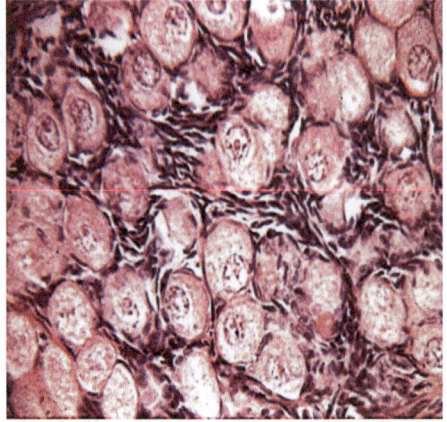

**Follicular (granulosa) cells**

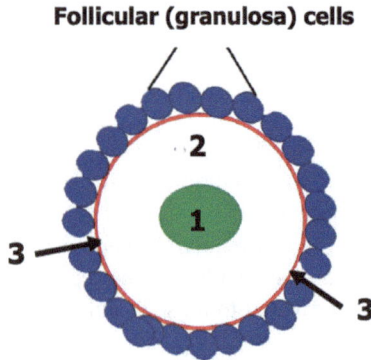

**Fig. 3.2** Drawing of a primary follicle. The oocyte is surrounded by cuboidal follicular cells. A small perivitelline space can be seen between the outer part of the oocyte cell membrane (3) and the inner one of the follicular cells. 1: Nucleus. 2: cytoplasm

The "selected" primordial follicles progress to primary follicles, which have now cuboidal follicular cells (Fig. 3.2).

The follicle cells proliferate, and when they form a 2 to 3, the follicles are named as secondary (Fig. 3.2). When the cells in the internal layer separate, leaving a space known as the antrum, the whole structure is called tertiary follicle. A thick layer known as the zona pellucida (ZP, secreted by the follicular cells) can be seen surrounding the vitelline membrane. This is the cytoplasmic membrane of the oocyte. A space is left between the internal part of the ZP, known as the perivitelline space, which is not identifiable with the light microscope (Figs. 3.3 and 3.4).

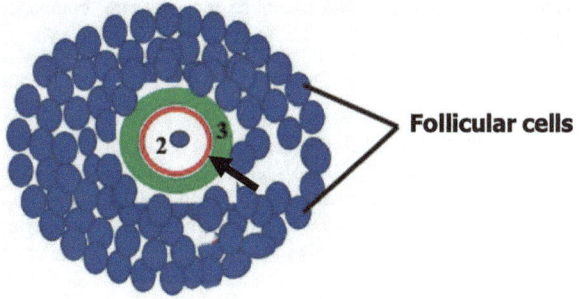

**Fig. 3.3** Scheme of an early secondary follicle. The antrum is being generated by separation of some follicular cells. 2: cytoplasm. Arrow: oocyte's cell membrane; 3: zona pellucida

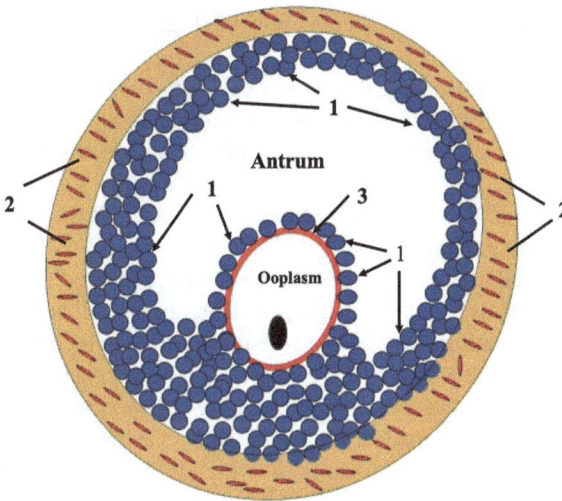

**Fig. 3.4** Schematic representation of a tertiary follicle in the process of maturing. The oocyte is surrounded by the vitelline membrane (not seen) and the zona pellucida (3). The latter is covered by some of the cumulus oophorus, a group of follicular cells (1). The thecal layer (2) encircled the follicle

Secondary follicle formation is independent from FSH (Cattanach et al. 1977) and dependency is acquired when follicles have 5 mm in diameter. The action of FSH receptors in follicular cells is mediated by insulin growth factors receptors (Monget and Bondy 2000).

During the prepubertal period, while tertiary follicles grow, they do not reach the 10 mm diameter typical of preovulatory follicles observed before ovulation.

Some of the follicular cells appear around the zona pellucida. They are collectively known as the cumulus oophorus (Fig. 3.4).

**Table 3.1** When the two meiotic divisions are completed, the tertiary oocyte has 1n chromosome content

|                    | 2n       | 4n             | 2n               | n                   |
|--------------------|----------|----------------|------------------|---------------------|
| Ovary              | Oogonium | Primary oocyte | Secondary oocyte | Tertiary oocyte     |
| Perivitelline space |         |                | First polar body | Secondary polar body |

During meiotic one, the ADN content in the oogonium is duplicated to 4n. After the first meiotic division is carried out at ovulation, the secondary oocyte has 2n and, finally, fertilization activates the secondary meiotic division to end up, and the tertiary oocyte results with 1n. The polar bodies are extruded to the perivitelline space with the correspondent n content, as shown; n: chromosome content

The transforming growth factor β and FSH are important in the development of follicular cells and hence for follicular growth (Ingman and Robertson 2009; Li et al. 2022).

Secondary follicle formation is independent from FSH (Cattanach et al. 1977) and dependency is acquired when follicles are 5 mm in diameter. The action of FSH receptors in follicular cells is mediated by insulin growth factors receptors (Monget and Bondy 2000).

Montazeri et al. (2022) found that cumulus oophorus AMH and FSH receptors are involved in oocyte maturation.

It is useful to recall that primary oocytes are detained in the diplotene stage of the first meiotic division, as explained in Chap. 1. The process will be resumed at ovulation, and DNA duplication occurs during the first meiosis, which is equivalent to 4n chromosome content. When that division ends, 2n will be excluded in the first polar body, which is extruded into the perivitelline space. The second meiosis occurs with fertilization; then, the chromosome content of the oocyte becomes 1n, with the other 1n being excluded in the second polar body, which is also located at the abovementioned site.

Hence, the oocyte will have 1n chromosome and given that the spermatozoid is also haploid, the 2n constitution of the adult mammal will be re-established (Table 3.1).

In assisted reproduction programs, it has been difficult to obtain good results, when activating fetal ovarian follicles which could mean that sufficient knowledge of molecular related events is lacking (Montazeri et al. 2022).

Before ovulation, the theca cells activate LH receptors, as to elaborate and secrete androgens, which will attach to granulosa cells receptors and are thus converted by the action of aromatase, in estrogens, mainly 17β estradiol (Fig. 3.5).

After ovulation, the thecal and follicular cells undergo molecular changes, leading to the expression of LH receptors and subsequent production and secretion of progesterone (see Chap. 7).

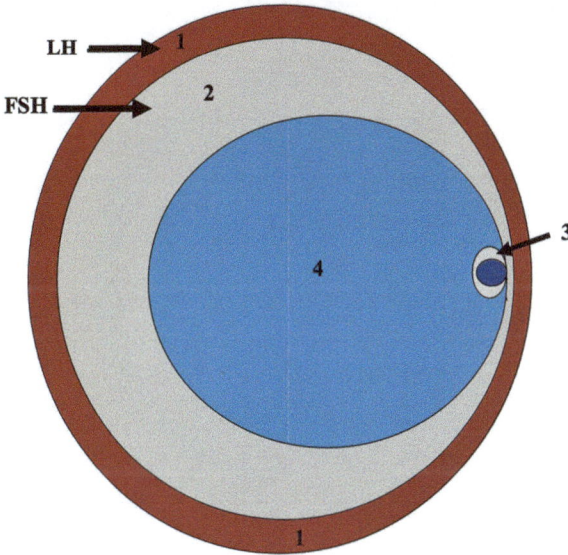

**Fig. 3.5** Presumptive tertiary follicle. Androgen LH receptors are expressed by theca cells (1), as to elaborate androgens. These molecules pass to the follicle cells (2) which transform androgens in estrogens. (3) Cumulus cells. (4) Antrum

# References

Cattanach BM, Iddon CA, Charlton HM et al (1977) Gonadotrophin-releasing hormone deficiency in a mutant mouse with kypogonadism. Nature Lond 269:338–340

Edson MA, Nagaraja AK, Matzuk MM (2009) The mammalian ovary from genesis to revelation. Endocr Rev 30(6):624–712

Ford EA, Beckett E, Roman S et al (2020) Advances in human primordial follicle activation and premature ovarian insufficiency. Reproduction 159(1):R15–R29

Ingman WV, Robertson SA (2009) The essential roles of TGF β1 in reproduction. Cytokine Growth Factor Rev 20(3):233–239

Li Q, Huo Y, Wang S et al (2022) TGF-β1 regulates the lncRNA transcriptome of ovarian granulosa cells in a transcription activity-dependent manner. Cell Prolif 56(1):e13336. https://doi.org/10.1111/cpr.13336

Monget O, Bondy C (2000) Importance of the IGF system in early folliculogenesis. Mol Cell Endocrinol 163(1,2):89–93

Montazeri F, Kalantar SM, Fesahat F et al (2022) Association between cumulus cells—mRNA levels of AMHR2 and FSHR with oocyte maturity. Middle East Fertil Soc J 27:26

Yang MH, Cushman RA, Fortune JE (2017) Anti-Müllerian hormone inhibits activation and growth of bovine ovarian follicles *in vitro* and is localized to growing follicles. Mol Hum Reprod 23(5):282–291

# Chapter 4
# Puberty: Acquisition of Ovulatory Capacity

**Abstract** Puberty means acquisition of reproductive capacity (ovulation), and its outcome depends on age, environmental, nutritional, and genetic factors. The comfort of herds is important in this context, as they collectively trigger the cessation of the existing hypothalamic blocking characteristic of the prepubertal period, which is related to ovarian function. Many explanations of this have arisen from research findings and relevant ones are presented in this chapter. Processes are illustrated and some possible interactive mechanisms are outlined.

**Keywords** Puberty · Ovulation · GnRH · LH · 17ß estradiol · FSH · Endorphin β · Galanin · TGF · Dopamine · Neuroregulin · Kisspeptin · Neuropeptide Y · Aspartate · Leptin · Thymulin · Amino acids · Hypothalamus · Hypophysis

Puberty means acquisition of reproductive capacity and its outcome depends on age, environmental, photoperiod, nutritional, and genetic factors. The wellbeing of herds is important in this context, as they collectively trigger the cessation of the existing hypothalamic blocking characteristic of the prepubertal period, which is related to ovarian function. Many explanations of this have arisen from research findings and relevant ones are presented in this chapter.

The studies carried out comprise the activation of the hypothalamus by neurotransmitters, amino acids, hormones, growth-promoting molecules, and various other substances secreted by the astrocytes and the thymus.

Before regular and cyclic ovarian function is established, there occurs a period of stabilization, which mainly includes ovulation and irregular chronology of ovarian cyclicity. The onset of puberty signs, such as male acceptance, increased secretion of vaginal glands, and changes in behavior, varies among different genotypes (Schillo et al. 1983, 1992).

The endocrinological sign about initiation of mature ovarian activity is the activation of GnRH secretion and subsequent increase in LH and FSH secretion. Neurotransmitters and modulators carry on information originated in environmental clues, metabolic products, energy resources, and body growth (Ebling 2005).

Evidence has been provided about a gradual disappearance of the blocking effect on GnRH activity (Levasseur 1977; Kiser et al. 1981; Day et al. 1984). Follicle growth enhances low levels of ovarian 17ß estradiol production, which provokes the appearance of external signs of the initiation of puberty (Evans et al. 1994).

At first, some outbursts in LH secretion can be detected, although they are not consistent in amount or regularity, to guarantee normal ovarian activity.

The bovine young female exposure to extended light periods can accelerate puberty appearance (Mezzadra et al. 1993; Rius and Dahl 2005). The pheromones contained in male's urine stimulate LH and FSH secretion (González-Padilla et al. 1975; Izard and Vandenbergh 1982) through the connection of olfactory neurons and the hypothalamus, as explained in Chap. 1.

As the ovarian follicles grow, estrogens levels in the blood increase. Likewise LH one and increasing outbursts ("peaks") of its secretion, characterized the main endocrine pattern of puberty. When the first ovulation is near, the number of estradiol receptors diminishes in the anterior and basal hypothalamic nuclei (Kinder et al. 1987), which could be one trigger in this context. In the arcuate nucleus, the amount of neuropeptide Y, a GnRH inhibitor, also goes down before puberty in monkeys. Endorphin β, a LH inhibitor, diminishes (Kinder et al. 1987; Wolfe et al. 1992; Genezzani et al. 2000; Honaramooz et al. 2000; Fig. 4.1).

Corticotrophin releasing factor diminishes at the onset of puberty; it could inhibit GnRH action in the hypothalamus during the prepubertal phase (Kinsey-Jones et al. 2010).

Galanin, a neuropeptide of wide distribution in the nervous system, is a candidate to participate in the mechanisms which trigger puberty commencement, although results in this topic are contradictory (Genezzani et al. 2000).

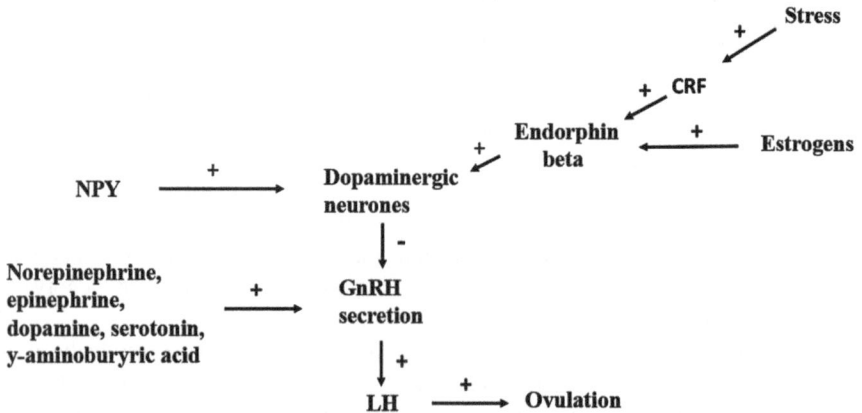

**Fig. 4.1** Some inhibitory (−) and promoting (+) mechanisms on LH secretion, which affect puberty onset. CRF: Adrenocorticotrophic hormone releasing factor. NPY: neuropeptide Y. LH: luteotrophic hormone. (Based on information presented by Kinder et al. (1987); Wolfe et al. (1992); Feleder et al. (1999); Genezzani et al. (2000))

The transforming growth factor (TGF) and neuregulin produced in astrocytes stimulate GnRH secretion. Since these cells have receptors for those molecules, signals between astrocytes are a way to controlling GnRH secretion (Ojeda et al. 2010).

TGFα activates epidermal growth factor receptors and also, another receptors for neuregulin, thus allowing liberation of prostaglandin E-2. This exerts a positive signal to stimulate GnRH secretion (Ma et al. 1994; Ojeda and Ma 1998; Fig. 4.2).

Before puberty, a light progesterone serum level rise was recorded, from 300 pg/ml to 1–3 ng/ml (González-Padilla et al. 1975). In another study, more than one increased levels were recorded, but they do not follow a pattern (Grajales et al. 2006). A possible manner of detecting the first ovulation at the onset of puberty is the detection of progesterone blood levels above 1 ng/ml (Jones et al. 1989).

These findings coincide with the findings of small completely nonfunctional corpora lutea, probably with a diminished capacity to express LH receptors (Berardinelli et al. 1979; Gregson et al. 2016).

Ovarian steroids stimulate the secretion of growth hormone (GH) and the insulin-like growth factor 1 (IGF1), leading to increases release of GnRH (Hull and Harvey 2014; Dosouto et al. 2019).

Receptor LH low number in the ovaries could be related to anovulation during the prepubertal period (Kinder et al. 1987).

Furthermore, the concourse of some amino acids has been proposed as GnRH inhibitors. For instance, aspartate receptors must be activated to permit GnRH

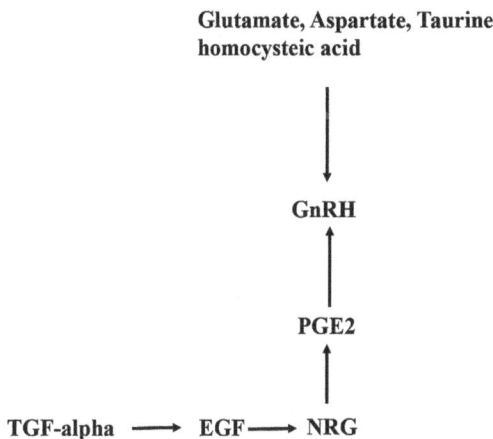

**Glutamate, Aspartate, Taurine homocysteic acid**

↓

**GnRH**

↑

**PGE2**

↑

**TGF-alpha** ⟶ **EGF** ⟶ **NRG**

**Fig. 4.2** Activation of EGF receptors by the stimulus of TGF-alpha or another molecules from astrocyte origin which attach to NRG, permits PGE-2 release, a GnRH stimulant. 1,5 GnRH metabolite controls GnRH secretion. All arrows represent positive effects. EGF: epidermal growth factor; GnRH: Gonadotrophin releasing hormone. PGE2: prostaglandin E2. NRG: neuregulin; TGF: transforming growth factor. Ma et al. (1994); Bourguignon et al. (1994); Ojeda and Ma (1998); Feleder et al. (1999)

(Bourguignon et al. 1994). Also, glutamate and taurine receptors must be activated to permit GnRH (Feleder et al. 1999).

Likewise, the vaso-active intestinal peptide is produced in the suprachiasmatic hypothalamic nucleus (Ebling 2005). The kisspeptin receptor GPR54 is located near the GnRH neurons. Shahab et al. (2005) proposed kisspeptin (a neurotransmitter) as another candidate to modulate GnRH.

In the prepubertal stage, gliogenesis and synaptogenesis are active processes, which appeared to be important in the development of GnRH neurons. Also, there is an increment of n-methyl-aspartate receptors, which receive increasing amount of glutamatergic impulses, needed to activate GnRH neurons (Adams et al. 1999).

These cells receive an increasing amount of glutamaterrgic impulses, in virtue of an increment of n-methyl-aspartate receptors, which activates GnRH neurons (Adams et al. 1999).

In Angus females, an increment of IGF-1 and growth hormone serum levels was found (Jones et al. 1989, 1991). IGF-1 permits GnRH output in the hypothalamic median eminence (Hiney et al. 1996; Yamada et al. 1998). IGF-1 is also needed for the growth of ovarian follicles, which is essential for puberty to occur (Simpson et al. 1991; Armstrong et al. 1992; Schoppee et al. 1996; Fig. 4.3). Figure 4.4 shows various factors, both stimulatory and inhibitory, involved in GnRH secretion.

Reinforcement in this context are the studies carried out by García et al. (2002), who found that IGF-1 and insulin blood levels increase before the onset of puberty.

The pineal gland melatonin is inhibitory for liberation of GnRH. At puberty, its blood levels diminish (Vanecek 1999).

Leptine is a protein known as the satiety hormone because it restrings appetite when the animal is satisfied. Likewise, leptine accelerates the oxidative metabolism. Food restriction in prepubertal heifers results in low levels of leptin and insulin in serum. Leptin also enhances the pulse frequency of LH secretion (Amstalden et al. 2000), which might imply that a suitable nutritional status is a must for the onset of puberty to occur.

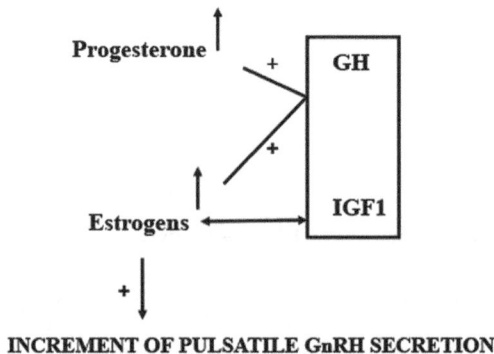

**Fig. 4.3** The sexual steroid serum levels before puberty appearance enhance growth hormone (GH) and insulin-like growth factor 1 (IGF1) production. GH and IGF1 could stimulate the frequency of GnRH pulsatile secretion. IGF1 possibly acts on estrogen elaboration (Belgorosky and Rivarola 1995; Schoppee et al. 1996)

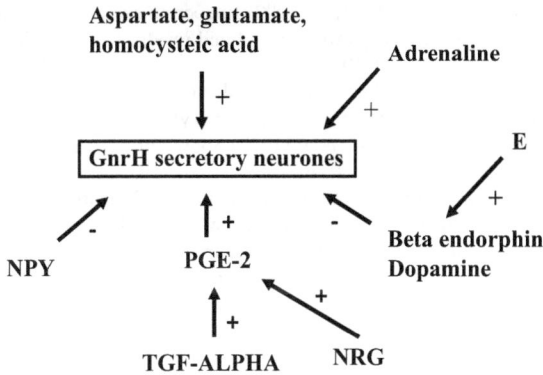

**Fig. 4.4** A collage of some inhibitory (−) or stimulatory (+) mechanisms on the GnRH secretion, which are active or inactive at puberty E = estrogens. NPY = neuropeptide Y. PGE-2 = prostaglandin E2. NRG = neuregulin. TGF= transforming growth factor. According to information gathered from various authors. Ma et al. (1994); Bourguignon et al. (1994); Ojeda and Ma (1998); Feleder et al. (1999)

Obese animals are infertile and have low serum levels of leptin. When they are treated with exogenous leptin, the reproductive activity is enhanced. Low blood leptin levels were associated with low FSH and LH production (Clarke and Henry 1999).

Leptin has direct action on GnRH, according to Zieba et al. (2004). Leptin blood levels augment before puberty (García et al. 2002).

According to the evidence given, leptin could have indirect and direct roles in GnRH secretion and ovarian follicle development.

Thymulin, produced by the thymus, could participate in the advent of puberty, as a positive factor to GnRh secretion (Hinojosa et al. 1999).

The multifactorial and intricate molecular events needed for the advent of puberty are not clear enough in spite of the important advances acquired. It seems that an astonishing synchrony of factors, as always, genetic and environmental, is needed for the onset of puberty.

# References

Adams MM, Flagg RA, Gore AC (1999) Follicular waves and circulating gonadotrophins in 8-month-old prepubertal heifers. Endocrinol 140:2288–2296

Amstalden M, García MR, Williams SW et al (2000) Leptin gene expression, circulating leptin and luteinizing hormone pulsatility are acutely responsive to short-term fasting in prepubertal heifers: relationships to circulating insulin and insulin-like growth factor I. Biol Reprod 63:127–133

Armstrong JD, Stanko RL, Cohick WS et al (1992) Endocrine events prior to puberty in heifers: role of somatotropin, insulin-like growth factor-I and insulin-like growth factor binding proteins. J Physiol Pharmacol 43:179–193

Belgorosky A, Rivarola MA (1995) Role of sex steroids in the mechanism of the onset of puberty. In: Bergada C and Moguilevsky JA (ed). Ares Serona symposia p 267

Berardinelli JG, Dailey RA, Butcher RL et al (1979) Source of progesterone prior to puberty in beef heifers. J Anim Sci 49:1276–1280

Bourguignon JP, Alvarez González ML, Gerard A et al (1994) Gonadotropin releasing hormone inhibitory autofeedback by subproducts antagonist at N-methyl-D-aspartate receptors: a model of autocrine regulation of peptide secretion. Endocrinol 134:1589–1592

Clarke IJ, Henry BA (1999) Leptin and reproduction. Rev Reprod 4(1):48–55

Day ML, Imakawa K, García-Winder M (1984) Endocrine mechanisms of puberty in heifers-estradiol negative feedback regulation of luteinizing hormone secretion. Biol Reprod 31:332–341

Dosouto C, Calf J, Polo A et al (2019) Growth hormone and reproduction: lessons learned from animal models and clinical trials. Front Endocrinol 10:404. https://doi.org/10.3389/fendo.2019.00404

Ebling FJP (2005) The neuroendocrine timing of puberty. Reproduction 129:675–683

Evans ACO, Adams GP, Rawlings NC (1994) Endocrine and ovarian follicular hanges leading up to the first ovulation in prepubertal heifers. J Reprod Fertil 100:187–194

Feleder C, Ginzburg M, Wuttke W et al (1999) GABAergic activation inhibits the hypothalamic-pituitary-ovaric axis and sexual development in the immature female rat. Associated changes in hypothalamic glutamatergic and taurinergic systems. Brain Res Dev Brain Res 116:151–157

García MR, Amstalden M, Williams SW et al (2002) Serum leptin and its adipose gene expression during pubertal development, the estrous cycle, and different seasons in cattle. J Anim Sci 80(8):2158–2167

Genezzani AR, Bernardi F, Monteleone P et al (2000) Neuropeptides, neurotransmitters, neurosteroids and the onset of puberty. Ann N Y Acad Sci 900:1–9

González-Padilla E, Wiltbank JN, Niswender GD (1975) Puberty in beef heifers. The interrelationship between pituitary, hypothalamic and ovarian hormones. J Anim Sci 40:1091–1104

Grajales H, Hernández A, Prieto E (2006) Determinación de parámetros reproductivos basados en los niveles de progesterona en novillas doble propósito en el trópico colombiano. Livest Res Rural Dev 18:144

Gregson E, Webb R, Sheldrick EL et al (2016) Molecular determinants of a competent bovine corpus luteum: first-vs-final wave dominant follicles. Reproduction 151(6):563–575

Hiney JK, Srivastava V, Nyberg CL et al (1996) Insulin-like growth factor I of peripheral origin acts centrally to accelerate the initiation of female puberty. Endocrinology 137(9):3717–3728

Hinojosa L, Chavira R, Domínguez R et al (1999) Effects of thymulin on spontaneous puberty and gonadotrophin-induced ovulation in prepubertal normal and hypothymic mice. J Endocrinol 163:255–260

Honaramooz A, Chandolia RK, Beard AP et al (2000) Opiodergic, dopaminergic, and adrenergic regulation of LH secretion in prepubertal heifers. J Reprod Fertil 119:207–215

Hull KL, Harvey S (2014) Growth hormone and reproduction: a review of endocrine and autocrine/paracrine interactions. Int J Endocrinol 2014:234014

Izard MK, Vandenbergh JG (1982) The effects of bull urine on puberty and calving date in crossbred beef heifers. J Anim Sci 55:1160–1168

Jones ER, Armstrong JD, Harvey RW (1989) Changes in metabolites, insulin and growth hormone prior to puberty in beef heifers. J Anim Sci 67(Supp 1):361

Jones ER, Armstrong JD, Harvey RW (1991) Changes in metabolites, metabolic hormones and luteinizing hormone before puberty in Angus, Bradford, Charolaise and Simmental heifers. J Anim Sci 69:1607–1615

Kinder JE, Day ML, Kittok RJ (1987) Endocrinology of puberty in cows and ewes. J Reprod Fertil Suppl 34:167–186

Kinsey-Jones JS, Li XF, Knox AMI et al (2010) Corticotrophin-releasing factor alters the timing of puberty in the female rat. J Neuroendocrinol 22(2):102–109

Kiser TE, Kraeling RR, Chapman JD (1981) Luteinizing hormone secretions before and after ovariectomy in prepubertal and pubertal beef heifers. J Anim Sci 53:1545–1550

Levasseur MC (1977) Thoughts on puberty: Initation of gonadotropic function. Ann Biol Anim Biochem Biophys 17:345–361

Ma YJ, Hill DF, Junier MP et al (1994) Expression of epidermal growth factor receptor changes in the hypothalamus during the onset of female puberty. Mol Cell Neurosci 5:246–262

Mezzadra C, Homse A, Sampedro D et al (1993) Pubertal traits and seasonal variation of the sexual activity in Brahman, Hereford ans crossbred heifers. Theriogenology 40:987–996

Ojeda SR, Ma YJ (1998) Epidermal growth factor tyrosine kinase receptors and the neuroendocrine control of mammalian puberty. Mol Cell Endocrinol 140(1–2):101–106

Ojeda SR, Lomniczi A, Sandau U (2010) Contribution of glial-neuronal interactions to the neuroendocrine control of female puberty. Eur J Neurosci 32:2003–2010

Rius AG, Dahl GE (2005) Exposure to long-day photoperiod prepubertally may increase milk yield in first lactation cows. J Dairy Sci 89:2080–2083

Schillo KK, Hansen L, Kamwanja D et al (1983) Influence of season on sexual development in heifers: age at puberty as related to growth and serum concentrations of gonadotropins prolactin, thyroxine and progesterone. Biol Reprod 28(2):329–341

Schillo KK, Hall J, Hileman S (1992) Effects of nutrition and season on the onset of puberty in the beef heifer. J Anim Sci 70:3994–4005

Schoppee PD, Armstrong JD, Harvey RW et al (1996) Immunization against growth hormone releasing factor or chronic feed restriction initiated at 3.5 months of age reduces ovarian response to pulsatile administration of gonadotropin-releasing hormone at 6 months of age and delays onset of puberty in heifers. Biol Reprod 55(1):87–98

Shahab M, Mastronardi C, Seminara SB et al (2005) Increased hypothalamic GPR54 signaling: a potential mechanism for initiation of puberty in primates. PNAS 102:2129–2134

Simpson RB, Armstrong JD, Harvey RW et al (1991) Effect of active immunization against growth hormone realising factor on growth and onset of puberty in beef heifers. J Anim Sci 69:4914–4924

Vanecek J (1999) Inhibitory effect of melatonin on GnRH-induced LH release. Rev Reprod 4:67–72

Wolfe MW, Robertson MS, Stumpf TT et al (1992) Modulation of luteinizing hormone and follicle-stimulating hormone in circulation by interactions between endogenous opioids and oestradiol during the peripuberal period of heifers. J Reprod Fertil 96:165–174

Yamada M, Hasegawa T, Hasegawa Y (1998) Increase in free insulin-like growth factor-I levels in precocious and normal puberty. Endocr J 45(3):407–412

Zieba DA, Amstalden M, Morton S et al (2004) Regulatory roles of leptin at the hypothalamic-hypophyseal axis before and after sexual maturation in cattle. Biol Reprod 71:804–812

# Chapter 5
# The Ovarian Cycle (Estrous Cycle) and Follicular Growth

**Abstract** The estrous cycle in cows is discussed in the context of environmental influencers, together with variations in their hormone levels, enhancers, and any inhibitors of follicle growth and maturation. A reasonable number of examples are available.

**Keywords** Estrus · 17β estradiol · Ovary · Environment · LH · Progesterone · Morphogenetic bone proteins · Inhibin · Differentiation growth factor 9 · Angiogenesis · Activin · Follicular waves · Insulin-like growth factors · Anti-Müllerian hormone · Hormone interactions

## General Overview

Estrus refers to the occurrence of external signs of sexual receptivity exhibited by a female animal. The time that elapses between the appearance of one estrus and that of the next one has been traditionally defined as the estrous cycle (EC). Given that ovulation and estrus are repetitive and cyclical, the EC is frequently known as the ovarian cycle.

This cycle, in the absence of pregnancy, lasts for 18–23 days in the majority of cattle, when ovarian activity becomes regular after puberty. In this study, the cycle was divided into two phases according to which hormone was predominant, thus yielding an estrogenic phase and a luteal phase, when blood progesterone levels increase (Fig. 5.1).

The majority of female animals have two follicular growth waves during the ovarian cycle. At the end of the first one, ovulation does not occur, because no follicles reach sufficient size, and therefore, only a limited amount of estrogen is produced. In the second wave, only one follicle develops, which is called an ovulatory. The duration of these two-wave cycles is between 18 and 21 days, and the duration of cycles featuring three waves is 25 or more (Sirois and Fortune 1988; Kopf et al. 1989; Adams et al. 1992).

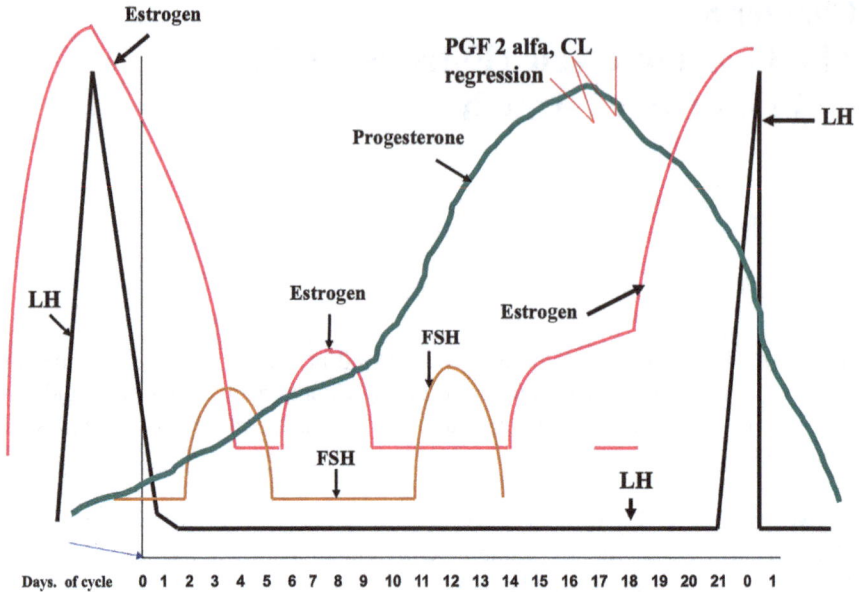

**Fig. 5.1** A sample representation of the serum levels of compromised hormones in the ovarian cycle of a cow. Real values and fluctuations are not shown. Before ovulation, which takes place during the first or second day of the cycle, estrogen levels rise according to the growth of the dominant follicle. That is accompanied by an abrupt elevation ("peak") in LH blood levels. After ovulation, a reduction in LH levels occurs, and these levels remain low until the end of the cycle. FSH enhances the development of nonovulatory ovarian follicles, which undergo two waves of growth. The second wave features a dominant follicle, the ovulatory one. Progesterone reaches its maximum level around the 16th day of the cycle. If a functional conceptus is absent, the corpus luteum regresses via the action of prostaglandin F2 alpha, and consequently, the progesterone levels decrease

Environmental conditions, both natural and imposed, affect reproduction phenomena. Nonadapted cattle genotypes show some degree of dysfunction when they are exposed to tropical environments without climatic seasons; here, only rainy and dry/semidry periods can be observed. Also, photoperiods are different from those in seasonal areas in the northern or southern hemisphere. For instance, in the tropics, variation in the duration of light over 24 hours is, at the most, 1 hour, whereas in seasonal countries, it can be 7 hours or many more. Similar variations in recorded temperatures are present in the abovementioned zones. For some *Bos taurus*– derived cattle in the Colombian tropics, in addition to their susceptibility to contracting tropical diseases, the climatic variables do affect reproduction. In relatively well-adapted Holstein cows on the Bogotá plain, at 2600 m above sea level, temperatures range between 7 and 19 °C at 75% humidity. Here, estrous appeared between 4:00 pm and 10:00 am the next day, when sunlight and the temperature were both in the low ranges (Cardozo et al. 1994). In another study (Vejarano and Hernández 2021), this time on Brahman and Romosinuano cows with different degrees of adaptation (where the creole Romosinuano of Spanish origin were the best adapted) to a dry tropical forest climate (in Armero, Tolima, Colombia; at 280 m above sea level; at temperatures ranging from 24 to 34 °C; and under high humidity), the durations of the estrous cycle were between 19.9 ± 1.6 days for Brahman cows and 21.2 ± 1.69 days for Romosinuano cows.

A heifer or a cow is "in estrus" (a period of sexual receptivity) when she mounts or allows mounting by other animals, changes her behavior, and eats less and when her vaginal vestibule becomes congested (indicated by redness) and somewhat edematous and vaginal mucous secretion is abundant. Estrous detection is economically important because it is key to program insemination given the limited life span of oocytes and spermatozoa. Ovulation takes place between 25 and 36 hours after signs of estrus are no longer evident. Spermatozoa and oocytes are viable for 2 days (Hansel and Convey 1983).

## Salient Endocrine Changes

Once follicle stimulating hormone (FSH) and luteinizing hormone (LH) patterns have been established during early puberty, ovarian cyclicity ensues. Before ovulation, LH links with corresponding thecal cell receptors. The result is androgen production. These hormones are converted to estrogens, mainly 17β, in follicular cells. During proestrus, estrogens that are produced by follicles and that will be ovulated feed back into the hypothalamus to induce increasing LH production and secretion by the basophilic cells during adenohypophysis. Estrogens prepare the cow and her reproductive tract to permit copulation and help spermatozoa to swim up the oviduct.

An LH secretion pattern is characterized by bursts ("peaks") of liberation.

When ovulation occurs, the follicle and thecal cells change their molecular machinery, together making up the corpus luteum, an intraovarian gland that starts producing progesterone (P), the principal hormone in pregnancy maintenance. On approximately the 16th day after fertilization, in the absence of a functional embryo, P blood levels decrease, mainly via the action of the uterine-produced prostaglandin F2α (Fig. 5.1; see Chaps. 7 and 8). High P blood levels block GnRH secretion.

Before each follicle's growth wave begins, during the estrous cycle, the level of FSH in blood can be detected (Fig. 5.1; Adams et al. 1992), which thereby indicates the level of estrogen in the blood. However, only in the last wave, the ovulatory one, does a follicle acquire sufficient capacity to be ovulated, as explained later on in this chapter. In the absence of a functional embryo in the uterus, estrogen serum levels increase, produced mainly by the ovulatory follicle, and a new cycle should then commence (Fig. 5.1).

In the preoptic hypothalamic area and the median eminence, some neurons produce an opioid known as endorphin-β (Eβ). Those cells are associated with GnRH-producing neurons. Eβ also inhibits GnRH secretion (Faletti et al. 1999). On the other hand, an adipose-stabilizing molecule, called neuropeptide Y (NPY),

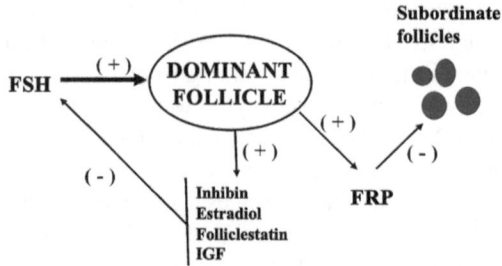

**Fig. 5.2** A possible inhibitory mechanism of FSH action occurs when FSH is no longer necessary for follicle development, because the dominant follicle has reached adequate growth. Inhibin, estradiol, folliculostatin, and some insulin-like growth factors (IGFs), which are produced by the dominant follicle, block FSH action. Another molecule, the so-called regulatory protein of follicle development (FRP) molecule, might act to impede the growth of subordinate follicles—ccording to the findings of Ireland (1987), Lucy et al. (1992), and Ginther et al. (2000). FSH: follicle-stimulating hormone

stimulates LH release by stimulating a GnRH linkage to the basophil cell receptors during adenohypophysis (Parker et al. 1991). LH secretion also responds to stimuli from periventricular cells in the preoptic area of the hypothalamus (Le et al. 1999; Fig. 5.2). Although the role of noradrenaline in controlling hypothalamic GnRH secretion has not been well defined, adrenergic receptors in the cytoplasm of GnRH-producing neurons have been encountered (Hosny and Jennes 1998).

# Development of Follicles

This process takes places thanks to interactions between inhibitors and promoters, like many other biological activities. In a wave of growth, from a big ovarian pool, a group of follicles are chosen to start growing (recruitment), but only one (dominant follicle) will be ovulated. Researchers have been devoted to trying to explain the biological bases of these phenomena.

In mammals, 3000 proteins have been identified as having presumptive roles in ovarian follicular growth, all of them dependent on gonadotrophins. Oocytes have been identified as participants by production of different active principles (Zhang et al. 2022).

Antral formation should ensue when follicles are 0.2–0.4 mm in diameter. The subordinated follicles will not reach a diameter of 10 mm; only the dominant follicle will (Lussier et al. 1987). The degeneration of antral follicles is initiated by generalized cell apoptosis, whereas autophagia accounts for follicular atresia (Meng et al. 2018).

Follicle growth has been characterized as having recruitment (follicles start growing), selection and dominance (of the ovulatory follicle) phases (Ireland 1987). In each wave of growth, three to six follicles (Sirois and Fortune 1988), or up to 24

(Hendriksen et al. 2003), begin increasing in diameter. Waves of growth occur between days 1 and 4 and between days 9 and 12 in the two waves of estrous cycles. When a third wave is exhibited, it appears at day 16 (Sirois and Fortune 1988; Ginther 1989; Evans et al. 1997). Three-wave cycles can last for 23 days (Ginther 1989).

One day before ovulation, some follicles are 4 mm in diameter. In the following days, one of them becomes dominant, but not ovulatory, and the remainder are subordinated (Ginther et al. 1996). In general, follicular growth proceeds at a regular rate until dominance has been established (Ginther et al. 1997, 1998, 1999; Kulick et al. 1999). At this point, the dominant follicle inhibits the growth of the subordinate follicles (Taylor and Rajamahendran 1991).

An increment in blood supply is hormonally controlled. Hence, dominance is determined by the augmentation of FSH and LH receptors and aromatase synthesis in the ovulatory follicle (Bao et al. 1997). In the recruitment phase, follicular growth depends on FSH and the continuation of development on LH (Webb et al. 1999).

Some molecules derived from oocyte, theca, and follicular cells stimulate or inhibit ovarian functions during follicular progression. Among them is growth differentiation factor 9 (GDF9), which is a member of transforming growth factor β (TGF-β) superfamily. The latter also includes morphogenetic bone proteins (BMP) and activins. BMP15 enhances the growth of follicular cells and non-FSH-dependent follicular progression (Kumar et al. 2017). BMP4 promotes the development and viability of primordial follicles in the ovary (Nilsson and Skinner 2003). Some researchers have proposed that the levels of saturated and unsaturated fatty acids should be in balance in the antral liquid because the unsaturated fatty acids negatively affect steroidogenesis in follicular cells, whereas the saturated fatty acids act in a contrary way (Baddela et al. 2022). The tyrosine kinase receptor and its ligand are important in follicle development as apoptosis suppressors and as stimuli for the migration and differentiation of ovarian cells and for follicular development (Driancourt et al. 2000).

Angiogenesis plays a prominent role in follicular development. Moreover, the endothelial growth factor (VEGF; Danforth et al. 2003) and fibroblastic growth factor (FGF) are key molecules in that process. They are regulated by the hypoxia inducible factor (HIF; Rizov et al. 2017).

Intrafollicular inhibin concentrations decrease during ovulatory follicle growth and increase during the follicular phase of the estrous cycle; 17β estradiol blood levels secreted by the ovulatory follicle increase, which inhibits the development of the subordinate follicles (Ireland 1987; Lucy et al. 1992; Fig. 5.2). Activin, FGFβ, and some morphogenetic proteins could prevent androgen synthesis in the thecal cells of subordinated follicles. Activin might also promote oocyte maturation (Knight and Glister 2006) and the inhibition of apoptosis in follicle cells (Shen and Zhu 2022). Bisphenol, an apoptosis enhancer, might play a role in the decline in the number of follicular cells in subordinated follicles (Sabry et al. 2023).

The dominant follicle produces estrogens and androgens that favor FSH function (Evans et al. 1997). Next, the dominant follicle's serum levels limit the development of subordinated follicles during the first wave of follicle growth. But the dominant follicle also synthesizes inhibin and somatostatin, antagonists of FSH action (Adams et al. 1992). Hence, a balance between promoters and inhibitors is established.

Esteroids and inhibin A both exert negative feedback on FSH secretion. Inhibin A is produced by follicle cells (Knight et al. 1989) and suppresses FSH synthesis during the early luteal phase (Turzillo and Fortune 1993). When FSH secretion reaches its maximum, its serum concentrations decrease over several days. Meanwhile, follicles continue growing until they have reached between 4 and 8.5 mm in diameter (Ginther et al. 1997, 1999). The dominant follicle's maximum growth is reached in 8 hours (Ginther et al. 1999). When the serum levels of estradiol are being secreted by the dominant follicle in greater amounts, they inhibit FSH synthesis (Barrett et al. 2006), as do insulin-like growth factors and inhibin (Ginther et al. 2000; Fig. 5.2).

## Inducers of Follicular Dominance

LH peaks of secretion are recognized as indicators of follicular dominance. They coincide with the synthesis of LH receptors in follicle cells (Sartori et al. 2001; Mihm et al. 2006; Marsters et al. 2015). Dominant follicle selection is promoted through the liberation of IGF-1, which is brought about by the transporters IGFBP-2 and IGFBP-4. This is allowed by some proteases. The same process is not present in subordinated follicles (Rivera et al. 2001). However, Zhou et al. (2003) argue that IGFBP-4 participates in the acquisition of luteal function, but not in the dominance event. Some influences that partially explain follicular dominance are presented in Fig. 5.3, as discussed by Knight and Glister (2006). Apelin, an adipokine, in vitro induces the hyperplasia of follicle cells from buffalo ovaries, a necessary condition for the development of the dominant follicle. Apelin might also reinforce the steroidogenesis of the IGF-1 stimulus (Shrokohalli et al. 2023).

In cows, follicles gain ovulatory capability in response to an LH wave, but this happens only when they are approximately 10 mm in diameter and have sufficient LH receptors (Sartori et al. 2001).

Nuclear vacuolization can be seen in follicular cells before the acquisition of dominance during proestrus. In the luteal phase, P exerts negative feedback on LH secretion. LH low- and high-frequency liberation pulses are characteristic of the time period preceding estrus, as opposed to the ample low-frequency pulses of the luteal phase (Shallenberger et al. 1985). The latter mode of secretion does not allow the dominant follicle to grow and is provoked by P-blocking action (Savio et al. 1993). This mechanism should apply to the inhibition of growth observed in the dominant follicle of the nonovulatory wave (Abeyawardene and Pope 1987). When insufficient P blood levels are detected, shorter estrous cycles (8–12 days) can be detected (Murphy et al. 1990; Savio et al. 1992). The same reason arguably explains the occurrence of irregular and brief estrous cycles before the establishment of regularity at the onset of puberty (see Chap. 4).

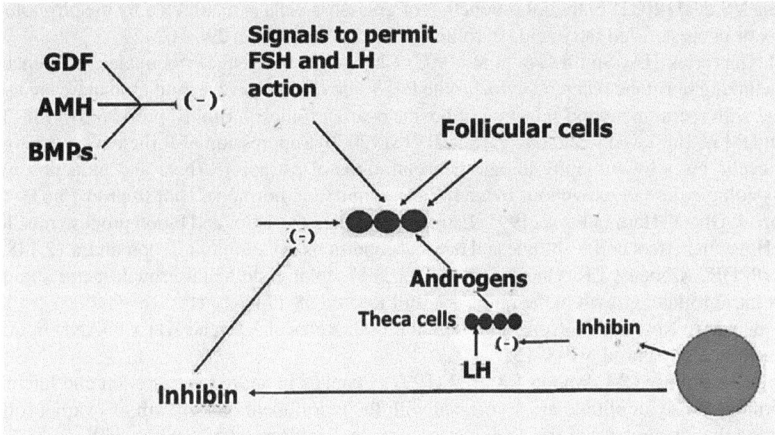

**Fig. 5.3** FSH, IGFs (insulin-like growth factors), estrogens, and androgens have positive effects on the follicular cells of subordinate follicles in that they promote follicular development. This process is inhibited by the negative action of the growth and differentiation factor (GDF), the anti-Müllerian hormone (AMH), bone morphogenetic proteins (BMPs), and the inhibin produced by the dominant follicle, which also blocks LH action—according to a description by Knight and Glister (2006)

A regulatory protein of follicular growth could be secreted by the dominant follicle to inhibit the development of subordinated follicles (Ireland 1987).

Between the LH/FSK peaks and ovulation, additional changes occur in the follicle, which are related to meiotic division and follicular rupture, and more changes occur in steroidogenesis, all for the follicle to gain the capacity for producing P as the main ovarian hormone after ovulation (Fortune et al. 2004).

Oocyte removal induces luteinization. Hence, this cell could play a role in follicular development (Matzuk et al. 2002).

The antiparamesonephric hormone (anti-Müllerian hormone, or AMH) is a marker of ovarian function in human beings. During follicle growth, the follicular cells secrete AMH until the antrum appears. Before ovulation, AMH secretion decreases (Moolhuijsen and Visser 2020).

In spite of the huge amount of information gathered about ovarian functions, many research questions remain.

# References

Abeyawardene SA, Pope GS (1987) The involvement of progesterone and luteinizing hormone in the termination of the post-ovulatory rise in plasma estradiol-17 beta concentrations in cattle. Anim Reprod Sci 15:27–36

Adams GP, Matteri RL, Ginther OJ (1992) Effect of progesterone on ovarian follicles, emergence of follicular waves and circulating follicle-stimulating hormone in heifers. J Reprod Fertil 95:627–640

Baddela VS et al (2022) Estradiol production of granulosa cells is unaffected by the physiological mix of non-esterified fatty acids in follicular fluid. J Biol Chem 298:102477

Bao B, Garverick HA, Smith GW et al (1997) Changes in messenger ribonucleic acid encoding luteinizing hormone receptor, cytochrome P450 side chain cleavage, and aromatase are associated with recruitment and selection of bovine ovarian follicles. Biol Reprod 56(5):1158–1168

Barrett DMW, Bartlewski PM, Duggavathi R et al (2006) Suppression of follicle wave emergence in cyclic ewes by supraphysiologic concentrations of estradiol-17beta and induction with a physiologic dose of exogenous ovine follicle-stimulating hormone. Biol Reprod 75:633–641

Cardozo J, Díaz F, Hernández A (1994) Estrous cycle characteristics and blood progesterone levels in Holstein heifers under altitude and tropical conditions in Colombia. Tropicultura 12:148–151

Danforth DR, Arbogast LK, Ghosh S et al (2003) Vascular endothelial growth factor stimulates preantral follicle growth in the rat ovary. Biol Reprod 68:1736–1741

Driancourt MA, Reynaud K, Cortvrindt R et al (2000) Roles of KIT and KIT LIGAND in ovarian function. Rev Reprod 5:143–152

Evans ACO, Komar CM, Wandji SA et al (1997) Changes in androgen secretion and luteinizing hormone pulse amplitude are associated with the recruitment and growth of ovarian follicles during the luteal phase of the bovine estrous cycle. Biol Reprod 57(2):394–340

Faletti AG, Mastronardi CA, Lomniczi A et al (1999) β-Endorphin blocks luteinizing hormone-releasing hormone release by inhibiting the nitricoxidergic pathway controlling its release. Proc Natl Acad Sci USA 96(4):1722–1726

Fortune JE, Rivera GM, Yang MY (2004) Follicular development: the role of the follicular microenvironment in selection of the dominant follicle. Anim Reprod Sci 82–83:109–126

Ginther O (1989) Follicular dynamics in heifers and mares. In: Society of Theriogenology. 1989. Proceedings of the annual meeting. Coeur d'Alene, Idaho, pp 1–11

Ginther OJ, Wiltbank MC, Fricke PM et al (1996) Selection of the dominant follicle in cattle. Biol Reprod 55:1187–1194

Ginther OJ, Kot K, Kulick LJ et al (1997) Emergence and deviation of follicles during the development of follicular waves in cattle. Theriogenology 48:75–87

Ginther OJ, Bergfelt DR, Kulick LJ et al (1998) Pulsatility of systemic FSH and LH concentrations during follicular-wave development in cattle. Theriogenology 50:507–519

Ginther OJ, Bergfelt DR, Kulick L et al (1999) Selection of the dominant follicle in cattle: establishment of follicle deviation in less than 8 hours through depression of FSH concentrations. Theriogenology 52:1079–1093

Ginther OJ, Bergfelt DR, Kulick LJ et al (2000) Selection of the dominant follicle in cattle: role of estradiol. Biol Reprod 63:383–389

Hansel W, Convey EM (1983) Physiology of the estrous cycle. J Anim Sci 57(Suppl 2):404–424

Hendriksen PJM, Gadella BM, Vos PLAM et al (2003) Follicular dynamics around the recruitment of the first follicular wave in the cow. Biol Reprod 69(6):2036–2044

Hosny S, Jennes L (1998) Identification of alpha(1B) adrenergic receptor protein in gonadotropin releasing hormone neurones of the female rat. J Neuroendocrinol 10(9):687–692

Ireland J (1987) Control of follicular growth and development. J Reprod Fertil Suppl 34:39–54

Knight PG, Glister C (2006) TGF-ß superfamily members and ovarian follicle development. Reproduction 132:191–206

Knight PG, Beard AJ, Wrathall JHM et al (1989) Evidence that the bovine ovary secretes large amounts of inhibin-subunit and its isolation from bovine follicular fluid. J Mol Endocrinol 2:189–200

Kopf L, Kastelic JP, Schallenberger E et al (1989) Ovarian follicular dynamics in heifers: test of two-wave hypothesis by ultrasonically monitoring individual follicles. Domest Anim Endocrinol 6(2):11–119

Kulick LJ, Kot K, Wiltbank MC et al (1999) Follicular and hormonal dynamics during the first follicular wave in heifers. Theriogenology 52:913–921

Kumar R, Alwani M, Kosta S et al (2017) BMP15 and GDF9 gene mutations in premature ovarian failure. J Reprod Infertil 18(1):185–189

Le WW, Beghorn KA, Rassnick S et al (1999) Periventricular preoptic area neurons coactivated with luteinizing hormone (LH)-realising hormone (LHRH) neurons at the time of the LH surge are LHRH afferents. Endocrinology 140(140):510–519

Lucy M, Savio J, Badinga L et al (1992) Factors that affect follicular dynamics in cattle. J Anim Sci 70:3615–3626

Lussier J, Matton P, Duffour J (1987) Growth rates of follicles in the ovary of the cow. J Reprod Fertil 81:301–307

Marsters P, Kendall N, Campbell B (2015) Pre-translational regulation of luteinizing hormone receptor in follicular somatic cells of cattle. Anim Reprod Sci 163:63–74

Matzuk MM, Burns KH, Viveiros MM et al (2002) Intercellular communication in the mammalian ovary: oocytes carry the conversation. Science 296:2178–2180

Meng L, Jan SZ, Hamer G et al (2018) Preantral follicular atresia occurs mainly through autophagy, while antral follicles degenerate mostly through apoptosis. Biol Reprod 99(4):853–863

Mihm M, Baker PJ, Ireland JL et al (2006) Molecularevidence that growth of dominant follicle involves a reduction in follicle stimulating hormone-dependence and an increase in luteinizing hormone-dependence. Biol Reprod 74:1051–1059

Moolhuijsen LM, Visser JA (2020) Anti-müllerian hormone and ovarian reserve: update on assessing ovarian function. J Clin Endocrinol Metab 105(11):3361–3373

Murphy M, Boland M, Roche J (1990) Pattern of follicular growth and resumption of ovarian activity in the postpartum beef suckler cow. J Reprod Fertil 90(2):523–533

Nilsson EE, Skinner MK (2003) Bone morphogenetic protein-4 acts as an ovarian follicle survival factor and promotes primordial follicle development. Biol Reprod 69(4):1265–1272

Parker SL, Kaira SP, Crowley WR (1991) Neuropeptide Y modulates the binding of a gonadotropin-releasing hormone (GnRH) analog to anterior pituitary GnRH receptor sites. Endocrinology 128(5):2309–2316

Rivera GM, Chandrasekher YA, Evans ACO et al (2001) A potential role for insulin-like growth factor binding protein-4 proteolysis in the establishment of ovarian follicular dominance in cattle. Bio Reprod 65(1):102–111

Rizov M, Andreeva P, Dimova I (2017) Molecular regulation and role of angiogenesis in reproduction. Taiwan J Obstet Gynecol 56(2):127–132

Sabry R, Williams M, LaMarre J et al (2023) Granulosa cells undergo BPA-induced apoptosis in a miR-21-independent manner. Exp Cell Res 427(1):113574

Sartori R, Fricke PM, Ferreira JC et al (2001) Follicular deviation and acquisition of ovulatory capacity in bovine follicles. Biol Reprod 65:1403–1409

Savio JD, Thatcher WW, Morris GR (1992) Terminal follicular development and fertility in cattle is regulated by concentration of plasma progesterone. 12th International Congress on Animal Reproduction, The Hague. Proceedings 2:999–1002

Savio JD, Thatcher WW, Morris GR et al (1993) Effects of induction of low plasma progesterone concentrations with a progesterone-releasing intravaginal device on follicular turnover and fertility in cattle. J Reprod Fertil 98(1):77–84

Shallenberger E, Schondorfer A, Walters D (1985) Gonadotrophins and ovarian steroids in cattle. I. Pulsatile changes of concentrations in the jugular vein throughout the oestrous cycle. Acta Endocrinol 108:312–321

Shen F, Zhu X (2022) Activin A reduces porcine granulosa cells apoptosis via ERβ-dependent ROS modulation. Vet Sci 9(12):704

Shrokohalli B, Zheng H-Y, Ma X-Y et al (2023) The effects of apelin on IGF1/FSH-induced steroidogenesis, proliferation, Bax expression, and total antioxidant capacity in granulosa cells of buffalo ovarian follicles. Vet Res Commun 47(3):1523–1533

Sirois J, Fortune J (1988) Ovarian follicular dynamics during the estrus cycle in heifers monitored by real-time ultrasonography. Biol Reprod 39:308–317

Taylor C, Rajamahendran R (1991) Follicular dynamics and corpues luteum growth and function in pregnany versus nonpregnant diary cows. J Dairy Sci 74:115–123

Turzillo A, Fortune J (1993) Suppression of the secondary FSH surge with bovine follicular fluid is associated with delayed ovarian follicular development in heifers. J Reprod Fertil 98:643–653

Vejarano A, Hernández A (2021) Grado de desarrollo del folículo preovulatorio y su relación con el tamaño del cuerpo lúteo y la producción de progesterona en las razas Romosinuano y Brahman. Rev Med Vet Zoot 68(1):52–65

Webb R, Campbell BK, Garveric HA et al (1999) Molecular mechanisms regulating follicular recruitment and selection. J Reprod Fertil 117:33–48

Zhang S, Mu L, Wang H et al (2022) Quantitative proteomic analysis uncovers protein-expression profiles during gonadotropin-dependent folliculogenesis in mice. Biol Reprod 108(3):479–491

Zhou J, Wang J, Penny D et al (2003) Insulin-like growth factor binding protein 4 expression parallels luteinizing hormone receptor expression and follicular luteinization in the primate ovary. Biol Reprod 69:22–29

# Chapter 6
# Ovulation

**Abstract** The anatomical changes of ovulation induced by hormones, cell secretions, and other molecules are presented, together with vascular and meiotic changes. Most processes are graphically illustrated.

**Keywords** GnRH · LH · Prostaglandin E · Prostaglandin F2α · Neutrophils · Eosinophils · Mast cells · Lymphocytes · HIF · VEGF · Thromboxane · FHL · Interleukins

Before ovulation, the P blood levels are in basal levels, and estrogens secreted by the ovulatory follicle stimulate GnRH secretion (Karsch 1987; Jaffe and Keys Jr 1974).

LH will command the ovum extrusion through the ovarian wall by vascular and proteolytic processes (Tsafriri and Reich 1999).

In the cow, the ovulatory follicle protrudes the ovarian wall and is more than 10 mm in diameter (Fig. 6.1).

Prostaglandins E and F2α have a role in the break out of the ovarian walls, composed of blood vessels, connective tissue, and celomic epithelium. Localized ischemia and a hypoxic environment facilitate ovulation (Magness et al. 1983). Migration of neutrophils, eosinophils, mast cells, and lymphocytes to the ovulatory zone plays a role (Cavender and Murdoch 1988; Alrabiah et al. 2021). Some of those cells liberate interleukins (Tsafriri and Reich 1999).

The first meiotic division ends at ovulation. Separation of the cumulus cells from the main pool of follicular cells is due to diminished vascular supply (Hirshfield 1991).

The oocyte leaves the ovary surrounded by cumulus follicle cells (Figs. 6.2 and 6.3). The proteolytic enzymes acting during ovulation, like plasminogen activator and collagenase, might be secreted by the celomic epithelium (Brännström et al. 1988; Espey and Lipner 1994; Fig. 6.4). This can be induced by chorionic gonadotrophic hormone, which has a similar LH action.

When ovulation is about to happen, there is an increment in blood supply to nonovulatory zones, in contrast with the already mentioned ischemia in the area

A. Hernández, *Bovine Maternal Support and Embryo Survival*,
https://doi.org/10.1007/978-3-031-62391-2_6

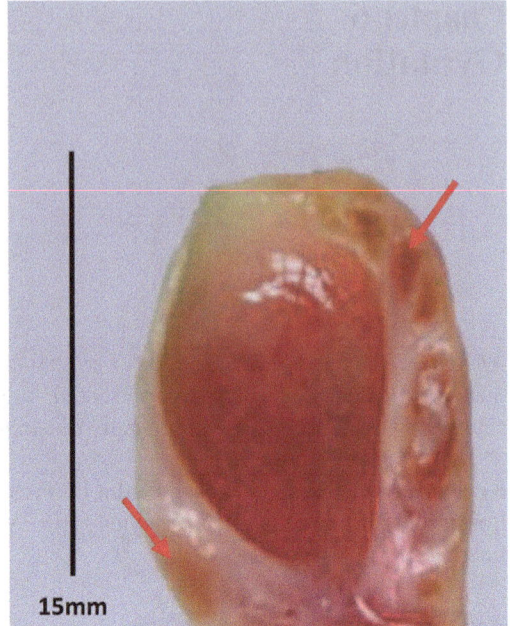

**Fig. 6.1** Ovary, cow. Ovulatory follicle. The wall was partially removed as to expose the follicle. Arrows: subordinate follicles. (Dr Álvaro Vejarano with permission. Modified)

15mm

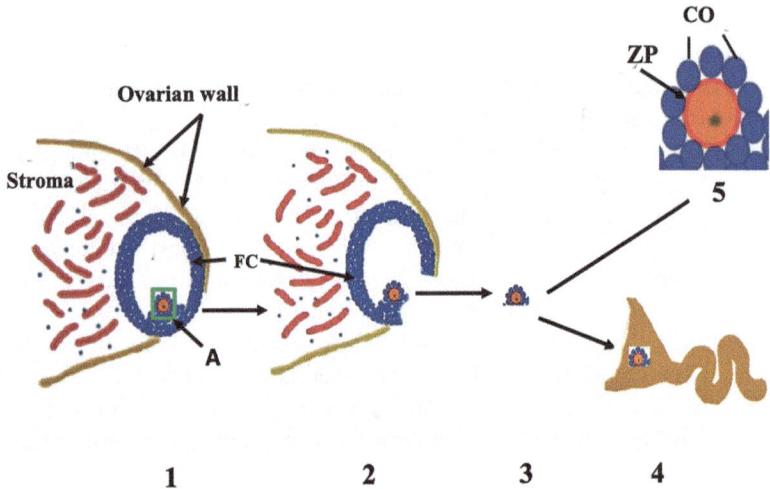

**Fig. 6.2** Sequential steps in ovulation (arrows). 1: lysis of the ovarian wall. 2–3: oocyte is expelled 2–3: oocyte release is surrounded by the zona pellucida and some cumulus cells. 4: the oviductal fimbria captures the ovulated structure (amplified in 5). A: zone with a higher level of hypoxia and less vascular supply. FA: follicular antrum. FC: follicular cells. ZP: zona pellucida. CO: cumulus cells

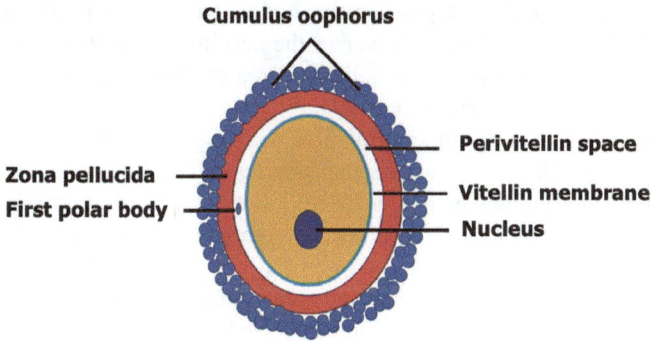

**Fig. 6.3** Representative drawing of an ovum as it is extruded from the ovary

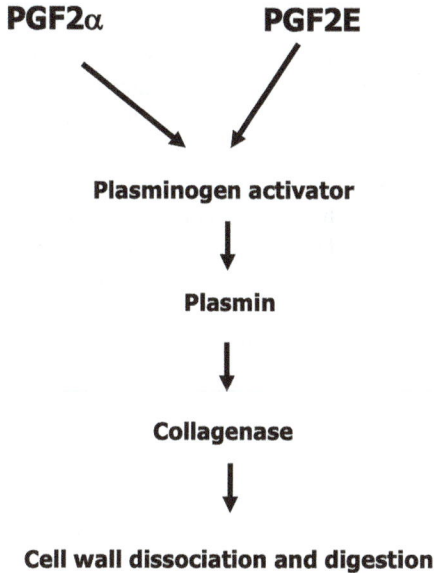

**Fig. 6.4** A proposed mechanism to explain cellular dissociation and death in the ovarian wall, according to Brännström et al. (1988)

where ovulation takes place, according to local biological demands (Shweiki et al. 1993; Neeman et al. 1997).

The hypoxic environment near the dominant follicle, but not in the ovulatory area, provokes HIF intervention. The latter sends signals for the expression of molecules in order to guarantee a physiological response. Among them, angiopoietins play a functional role (Fraser 2006). When the cumulus cells separate from the follicle and thecal cells, the vascular supply diminishes (Hirshfield 1991).

VEGF favors follicular cell proliferation and vascular permeability in the ovulatory follicle. Water diffusion is slower in the antrum than in the follicular cells. Hence, the volume of the ovulatory follicle increases, thus helping ovum extrusion (Shweiki et al. 1993; Neeman et al. 1997).

Interleukin 1 (IL-1) can induce ovulation, probably as a stimulus for hyaluronic acid synthesis, which is necessary for the cumulus cell changes. Also, IL-1 induces NO synthase and gelatinase activation, as well as prostaglandin synthesis. The interleukin 8 would activate heterophil accumulation and migration (Tsafriri and Reich 1999).

The activity of prostaglandins, thromboxanes, and leukotrienes is stimulated by LH. They stimulate smooth muscle contraction in the follicular wall and the vascular reaction (Murdoch and Myers 1983). PGF2α is essentially important in ovulation, since the inhibitors of cyclooxygenase block that event (Murdoch and Myers 1983; Bridges and Fortune 2007). However, anti-prostaglandins inhibit this route, but not completely (Li et al. 1991). The platelet-activating factor is a potent mediator in inflammation and also has a role in ovulation. Its action is under LH control (Li et al. 1991).

Nitric oxide could be involved in ovulation, causing vasodilatation. This would facilitate leukocyte diapedesis. Bradykinin, histamine, the renin-angiotensin complex, and free radicals are claimed to have a role in this context (Tsafriri and Reich 1999; Ben-Ami et al. 2006).

The first meiotic division, which ends during ovulation, in humans and rats, depends on the concourse of amphiregulin and epiregulin, which belong to the epidermal growth factor family. They are regulated by FSH and LH (Ben-Ami et al. 2006). Those substances can also activate oocyte maturity and follicular growth (Hsieh et al. 2009).

Cyclic adenosine monophosphate in high levels maintains the arrest of oocyte's meiosis, and, at low levels, meiosis is reinitiated in a LH-dependent manner (Conti et al. 1998).

Vaknin et al. (2001) found that a sterol could activate meiosis in rat's ovaries during the preovulatory period.

In preparation for the success of the second meiotic process, two kinases, RAS and ERK1 and 2, are activated during LH surge in follicular cells (Fan et al. 2009).

FHL is a molecule composed by proteins Lin-1, Isl-1, and Mec-3, which, together with the epidermal growth factor receptor and the so-called Hippo/YAP (a growth promoter), functions in follicle cells to enhance growth (Wang et al. 2023).

In monkeys, if neurotensin is immunologically inhibited, ovulation is blocked (Campbell et al. 2022).

After ovulation, the celomic epithelium proliferates as to comply with tissue repair (Tan and Fleming 2004).

# References

Alrabiah NA, Evans ACO, Fahey AG et al (2021) Immunological aspects of ovarian follicle ovulation and corpus luteum formation in cattle. Reproduction 162:209–225

Ben-Ami I, Freimann S, Armon L et al (2006) Novel function of ovarian growth factors: combined studies by DNA microarray, biochemical and physiological approaches. Mol Hum Reprod 12:413–419

Brännström M, Woessner F, Koos R et al (1988) Inhibitors of mammalian tissue collagenase and metalloproteinases suppress ovulation in the perfused rat ovary. Endocrinology 122(5):1715–1721

Bridges PJ, Fortune JE (2007) Regulation, action and transport of prostaglandins during the periovulatory period in cattle. Mol Cell Endocrinol 263(1–2):1–9

Campbell GE, Bender HR, Parker GA et al (2022) Neurotensin: a novel mediator of ovulation? FASEB J 35(4):e21481

Cavender JL, Murdoch WJ (1988) Morphological studies of the microcirculation system of periovulatory ovine follicles. Biol Reprod 39:989–997

Conti M, Andersen CB, Richard FJ et al (1998) Role of cyclic nucleotide phosphodiesterases in resumption of meiosis. Mol Cell Endocrinol 145:9–12

Espey LL, Lipner H (1994) Ovulation. In: Knobil E, Neill JD (eds) The physiology of reproduction. Reven Press, New York, pp 725–780

Fan H-Y, Masayuki ZL, Shimada M et al (2009) MAPK3/1 (ERK1/2) in ovarian granulosa cells are essential for female fertility. Science 324:938–941

Fraser HM (2006) Regulation of the ovarian follicular vasculature. Reprod Biol Endocrinol 4:18

Hirshfield A (1991) Development of follicles in the mammalian ovaries. Int Rev Cytol 124:43–101

Hsieh M, Zamah AM, Conti M (2009) Epidermal growth factor-like growth factors in the follicular fluid: role in oocyte development and maturation. Semin Reprod Med 27:52–61

Jaffe RB, Keys WR Jr (1974) Estradiol augmentation of pituitary responsiveness to gonadotropin-releasing hormone in women. J Clin Endocrinol Metab 39:850–855

Karsch FJ (1987) Central actions of ovarian steroids in the feedback regulation of pulsatile secretion of luteinizing hormone. Annu Rev Physiol 49:365–382

Li XM, Sagawa N, Ihara Y et al (1991) The involvement of platelet activating factor in thrombpcytopenia and follicular rupture during gonadotropin-induced superovulation in immature rats. Endocrinology 129:3132–3138

Magness RR, Christenson RK, Ford SP (1983) Ovarian blood flow throughout the estrous cycle and early pregnancy in sows. Biol Reprod 28:1090–1096

Murdoch W, Myers D (1983) Effect of treatment of estrus ewes with indomethacin on the distribution of ovarian blood to the periovulatory follicle. Biol Reprod 29:1229–1232

Neeman M, Abramovitch R, Schiffenbauer YS et al (1997) Regulation of angiogenesis by hypoxic stress: from solid tumours to the ovarian follicle. Int J Exp Pathol 78:57–70

Shweiki D, Itin A, Neufeld GG et al (1993) Patterns and expression of vascular endothelial growth factor (VEGF) and VEGF receptors in mice suggest a role in hormonally regulated angiogenesis. J Clin Invest 91:2235–2243

Tan OL, Fleming JS (2004) Proliferating cell nuclear antigen immunoreactivity in the ovarian surface epithelium of mice of varying ages and total lifetime ovulation number following ovulation. Biol Reprod 71:1501–1507

Tsafriri A, Reich R (1999) Ovulation in mammals. Exp Clin Endocrinol Diabetes 107:1–11

Vaknin KM, Lazar S, Popliker M et al (2001) The role of meiosis activating sterols in rat oocyte maturation: effects of specific inhibitors and changes in the expression of lanosterol 14α-demethylase during the preovulatory period. Biol Reprod 64:299–309

Wang C, David JS, Sun H et al (2023) FHL2 deficiency impairs follicular development and fertility by attenuatin g EGF/EGFR/YAP signaling in ovarian granulosa cells. Cell Death Dis 14:239. https://doi.org/10.1038/s41419-023-05759-3

# Chapter 7
# Formation of the Corpus Luteum

**Abstract** The corpus luteum occupies the space left by the extruded ovum and some attached cumulus follicle cells. The non-cumulus follicular cells and the theca ones constitute the corpus luteum, a temporary endocrine unit, which mainly produces progesterone, as to maintain pregnancy. The inherent mechanism to provoke pertinent events of corpus luteum construction is presented in the light of current research findings. Processes and possible interactions are graphically illustrated.

**Keywords** Corpus luteum · LH · Progesterone · Regulatory protein of luteinization · Tyrosine kinase receptor · Serotonin · HIF · VEFG · Renin-angiotensin complex · Notch system · Apelin · Resistin · Leptin · Neutrophils

When the oocyte is extruded, the thecal and follicular cells, which are left behind, express LH receptors. They bind the hormone receptors and produce progesterone (P). Both types of cells are converted in an endocrine intraovarian gland, the corpus luteum (CL). This structure will remain functional (producing P) up to the 16th day of the estrous cycle, if a functional conceptus (the embryo and its membranes) is not present in the uterus, or throughout pregnancy, as the main hormone (Figs. 7.1, 7.2, and 7.3).

When the CL is being established, LH action is greatly inhibited by P action on the hypothalamic GnRH secretion. Follicular estrogens and inhibitory peptides, after ovulation, allow for a FSH liberation to initiate a new wave of follicular growth (Manns et al. 1984).

The breaking of the ovarian wall during ovulation causes hemorrhage, and the follicle collapses. This results in the appearance of a hemorrhagic zone in the site of ovulation (Niswender and Nett 1994).

The follicular cells are converted in the large luteal cells. The small ones are the previous thecal cells (Fig. 7.3). The thecal cells proliferate and undergo hypertrophy, whereas the follicular cells do not divide. They now lack the capability of producing androgens (Ursely and Lemaire 1979; Wiltbank et al. 1991). Follicle cells stop producing the P450 and P450 17α enzymes, necessary for elaborating

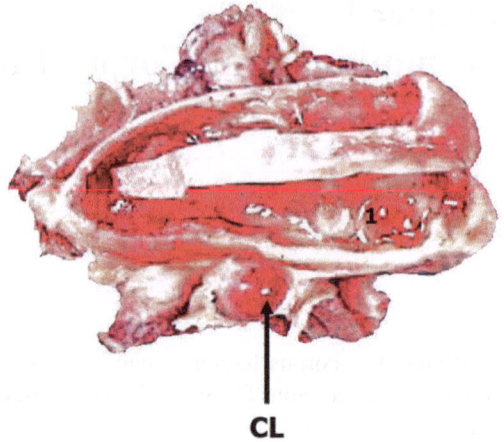

**Fig. 7.1** Corpus luteum (CL) in a pregnant ewe. 1: embryo

**CL**

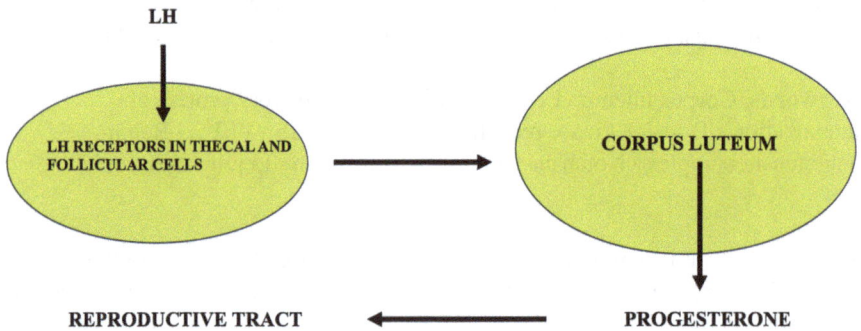

LH

LH RECEPTORS IN THECAL AND FOLLICULAR CELLS

CORPUS LUTEUM

REPRODUCTIVE TRACT

PROGESTERONE

**Fig. 7.2** Origin of the corpus luteum. The follicular and thecal cells after ovulation ensue progesterone production in the corpus luteum

estrogens (Murphy 2000). The regulatory protein of luteinization (StAR) is expressed in luteal cells (Orly and Stocco 1999).

The large luteal cells also produce relaxin (Fields and Fields 1996), an important hormone that helps in parturition, as well as growth factors (Miyamoto et al. 1992).

The luteal large cells produce about 80% of luteal P, in spite of exhibiting less LH receptors than small cells. P secretion is regulated by the number of occupied LH receptors (Chegini et al. 1991). LH concentrations are not high during the mid-luteal phase. The number of occupied receptors in luteal cells augments 40 times between days 2 and 10 of the ovarian cycle, and it remains elevated until the 14th day. Prolactin and the growth hormone (both of them produced in the acidophilic cells of the adenohypophysis) are important to maintain an appropriate P action. The key point to regulate P action is the transportation of cholesterol to the mitochondria (Niswender and Nett 1994; Juengel and Niswender 1999). The tyrosine kinase receptor KIT, expressed by small luteal cells, participates in CL formation (Spanel-Borowski et al. 2007).

**Fig. 7.3** Corpus luteum (mouse). Small and darker nuclei belong to cells, which originally are thecal cells. Big ones belong to luteal cells, which are the former follicular cells

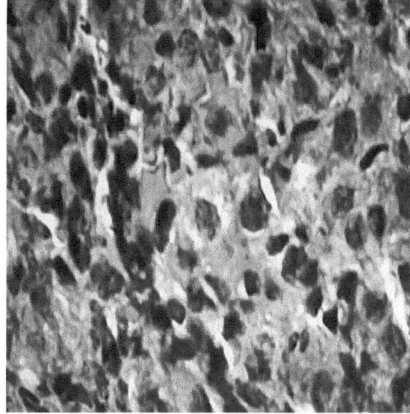

Adrenaline could stimulate P secretion after the tenth day of the estrous cycle (Payne and Cooke 1998). Also, serotonin appears to have the same effect (Payne et al. 1994).

The CL is refractory to the lytic PGF2α action during days 1–5 of the estrous cycle (see Fig. 5.1) (Goravanhally 2009).

CL formation needs an adequate blood supply (Shweiki et al. 1993), which is induced by hypoxia, the consequent HIF expression and then VEGF participation (Forsythe et al. 1996). The capillary density in CL is higher as compared to many organs (Reynolds and Redmer 1999). The control of vascular development is exerted by vasoinhibin-1 produced in the endothelium of the CL blood vessels (Shirasuna et al. 2012). The renin-angiotensin complex has also a role in angiogenesis (Davis et al. 2003). Prorenin, a renin receptor precursor, stimulates P secretion (Pereira et al. 2022).

The conserved Notch system affects proliferation and apoptosis, plays a role in CL formation in rats, since one of the ligand proteins of the Notch family, called DLL4, is blocked, CL viability is affected (Accialini et al. 2015).

Leptine, an adipokine, promotes P secretion, by stimulating growth hormone function. Also, it enhances apoptosis (Gregoraszczuk and Ptak 2005). Another lipokine, apelin, stimulates P production in pig's CL, at different stages of development (Różycka et al. 2020). The same action is exerted by resistin, another adipokine (Kurowska et al. 2021).

The presence of neutrophils in the CL might enhance P secretion and angiogenesis (Jiemtaweeboon et al. 2011).

# References

Accialini P, Hernández SF, Bas D et al (2015) A link between Notch and progesterone maintains the functionality of the rat corpus luteum. Reproduction 149:1–10

Chegini N, Lei M, Rao CV et al (1991) Cellular distribution and cycle phase dependency of gonadotropin and eicosanoid binding sites in bovine corpora lutea. Biol Reprod 45:506–513

Davis JS, Rueda BR, Spanel-Borowski K (2003) Microvascular endothelial cells of the corpus luteum. Reprod Biol Endocrinol 10(1):89

Fields MJ, Fields PA (1996) Morphological characteristics of the bovine corpus luteum during the estrous cycle and pregnancy. Theriogenology 45(7):1295–1325

Forsythe JA, Jiang BH, Iyer NV et al (1996) Activation of vascular endothelial growth factor gene transcription by hypoxia-inducible factor 1. Mol Cell Biol 16:4604–4613

Goravanhally MP (2009) Cellular mechanisms responsible for development of sensitivity of the bovine corpus luteum to prostaglandin F2 alpha. Dissertation, West Virginia University USA

Gregoraszczuk EL, Ptak A (2005) In vitro effect of leptin on growth hormone (GH)- and insulin-like growth factor-I (IGF-I)-stimulated progesterone secretion and apoptosis in developing and mature corpora lutea of pig ovaries. J Reprod Dev 51:727–733

Jiemtaweeboon P, Shirasuna K, Kobayashi A et al (2011) High number of neutrophils exists in the developing corpus luteum and promotes angiogenesis and progesterone secretion in the cow. Biol Reprod 85(Suppl_1):218

Juengel JL, Niswender GD (1999) Molecular regulation of luteal progesterone synthesis in domestic ruminants. J Reprod Fertil 117:193–205

Kurowska P, Sroka M, Dawid M et al (2021) Expression and role of resistin on steroid secretion in the porcine corpus luteum. Reproduction 162:237–248

Manns J, Niswender G, Braden T (1984) FSH receptors in the bovine corpus luteum. Theriogenology 22:321–328

Miyamoto A, Okuda K, Schweigert FJ et al (1992) Effects of basic fibroblast growth factor, transforming growth factor-beta and nerve growth factor on the secretory function of the bovine corpus luteum in vitro. J Endocrinol 135(1):103–114

Murphy BD (2000) Models of luteinization. Biol Reprod 63(1):2–11

Niswender GD, Nett TM (1994) Corpus luteum and its control in infraprimate species. In: Knobil E, Neill JD (eds) The physiology of reproduction. Raven Press, New York, pp 781–816

Orly J, Stocco DM (1999) The role of the steroidogenic acute regulatory (StAR) protein in female reproductive tissues. Horm Metab Res 31:389–398

Payne JH, Cooke RG (1998) Effect of adrenalin and propranolol on progesterone and oxytocin secretion in vivo during the caprine estrous cycle. Theriogenology 49(4):837–844

Payne JH, Vaillant C, Cooke RG (1994) Serotonin as a possible physiological regulator of ovine luteal function. Anim Reprod Sci 36(1–2):103–112

Pereira AMP, Da Rosa PR, Dos Santos J et al (2022) The influence of prorenin/(pro)renin receptor on progesterone secretion by the bovine corpus luteum. Anim Reprod Sci 241:106985

Reynolds LP, Redmer DA (1999) Growth and development of the corpus luteum. J Reprod Fertil Suppl 54:181–191

Rózycka M, Kurowska P, Grzesiak M et al (2020) Apelin and apelin receptor at different stages of corpus luteum development and effect of apelin on progesterone secretion and 3β-hydroxysteroid dehydrogenase (3β-HSD) in pigs. Anim Reprod Sci 192:251–260

Shirasuna K, Kobashi A, Nitta A et al (2012) Possible action of vasohibin-1 as an inhibitor in the regulation of vascularization of the bovine corpus luteum. Reproduction 143:491–500

Shweiki D, Itin A, Neufeld G et al (1993) Patterns of expression of vascular endothelial growth factor (VEGF) and VEGF receptors in mice suggest a role in hormonally regulated angiogenesis. J Clin Invest 9:2235–2243

Spanel-Borowski K, Sass K, Löffler S et al (2007) KIT receptor-positive cells in the bovine corpus luteum are primarily theca-derived small luteal cells. Reproduction 134(4):625–634

Ursely J, Lemaire P (1979) Varying response to luteinizing hormone of two luteal cell types isolated from bovine corpus luteum. J Endocrinol 83:303–310

Wiltbank M, Diskin M, Niswender G (1991) Differential actions of second messenger systems in the corpus luteum. J Reprod Fertil Suppl 43:65–75

# Chapter 8
# Luteal Regression

**Abstract** Since the corpus luteum (CL) is a transitory endocrine gland, a progesterone (P) provider, when gestation does not occur, it should disappear. It means that luteotrophic agents stop acting. The promoter molecules are to be replaced by lytic agents, to cause luteal regression. Many events are illustrated.

**Keywords** Regression · Prostaglandin F-2$\propto$ · GnRH · FSH · Oxytocin · Corpus luteum · Corpus albicans · Regulatory steroidogenic protein · Endothelin-1 · Gamma interferon · fas-L

Since the corpus luteum (CL) is a transitory endocrine gland, a progesterone (P) provider, when gestation does not occur, it should disappear. It means that luteotrophic agents stop acting. The promoter molecules are to be replaced by lytic agents, to cause luteal regression.

As already discussed, LH is a protagonist in luteal formation and functionality. At the end of the estrous cycle, the amplitude and frequency of LH change. The FSH action permits the development.

In a normal estrous cycle, the CL is gradually regressed after the 16th day. Prostaglandin F-2$\propto$ (PGF-2$\propto$) has been accepted as the main luteolytic agent. It is produced in the uterine horn, ipsilateral to the ovulation site (reviewed by McCracken et al. (1999) and Niswender et al. (2000)). PF2$\propto$ in ruminants goes through the uterine veins to the ovarian artery in a countercurrent way (Heap et al. 1985).

## The Mechanism of Luteal Regression

Progesterone receptors in the hypothalamus and uterus decline (down regulation) at the end of the progestational phase of the two waves-estrous cycle (days 17–19, diestrus). In consequence, GnRH will induce FSH secretion and estrogens from the dominant follicle, and oxytocin (in the hypothalamus and neurohypophysis) is

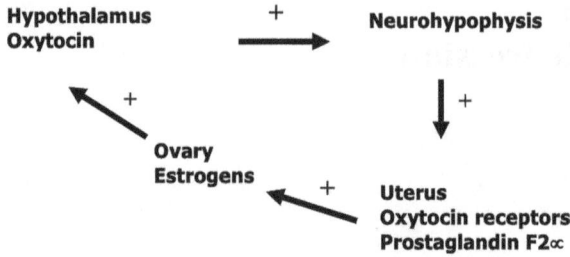

**Fig. 8.1** When progesterone blood levels diminish, oxytocin and PGF2 provoke corpus luteum regression. (Information taken from McCracken et al. (1996) and Silvia et al. (1991))

**Fig. 8.2** Corpus luteum regression. Oxytocin and estrogens stimulate PGF2α uterine secretion. PGF2α exerts positive feedback to enhance oxytocin secretion. Burns et al. (1997); Oliveira et al. (2022)

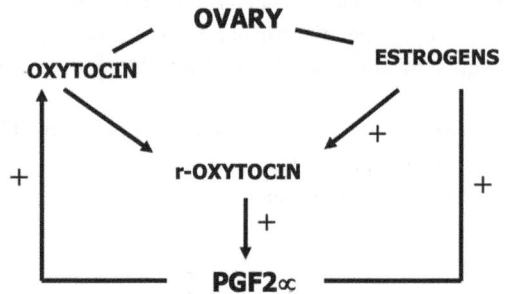

liberated and oxytocin receptors in the endometrium are synthesized. These receptors bind oxytocin and stimulate FGF2α production to cause luteal regression (McCracken et al. 1996; Silvia et al. 1991; Fig. 8.1). The PGF2α progressive secretion stimulates oxytocin liberation in the ovary, thus amplifying the process (Fuchs et al. 1990; Burns et al. 1997).

Ovarian oxytocin establishes a positive feedback with uterine PGF2α, about the 17th day of the estrous cycle. Estrogens from the ovulatory follicle stimulate the synthesis of oxytocin uterine receptors and the transcription factors associated with PGF2α synthesis in the uterus (Oliveira et al. 2022; Fig. 8.2).

Luteal regression comprehends a structural involution, which entails the development of connective tissue in the site previously occupied by the CL. That vestige is known as the corpus albicans (Skarzynski et al. 2008).

Prostaglandin E (PGE) is luteotropic, and it should be in balance with luteolytic PGF2α. The CL midlife span depends on that equilibrium. During luteolysis, PGE and its receptors are downregulated. Hence, PGF2α and its receptors predominate (Burns et al. 1997; Arosh et al. 2004). Twenty-four hours after P levels start to go down, PGF2α synthesis ascend to 300%, and there is a 50% increment in OX synthesis. Steroidogenesis is detained due to a lack in blood supply, thus affecting LH arrival to the luteal cells and LH receptors are downregulated. Also, cholesterol transport to internal mitochondrial membrane is shot down as a consequence of inhibition of the regulatory steroidogenic protein StAR (Diaz et al. 2002). Blood supply deficiency is a key event in luteal regression. At early stages of luteal regression, there is an acute increment of blood supply, possibly by an increment of nitric oxide, stimulated by PGF2α. However, 2–4 hours later, vasoconstriction takes place (Shirasuna 2010). Vasoconstriction appears to be mediated by endothelin-1 (ET-1); its levels go up (72%) during luteal decay and those of angiotensin 2 show an increment of 27.7%. Angiotensin 2 reinforces PGF2α action (Shirasuna et al. 2004).

Vasoconstriction leads to hypoxia and lack of nutrients including substrates to elaborate P and luteotropic factors. ET-1 promotes migration of leukocytes (mainly lymphocytes and macrophages) and macrophage cytokines' liberation (Penny et al. 1998). Tumor necrosis factor, gamma interferon, and Fas-L are also involved in luteolysis by apoptosis induction (Shaw and Britt 1995; Meidan et al. 1999). Also, necroptosis and autophagia can be involved (Jonczyk et al. 2019; Hojo et al. 2022). More molecules were proposed to be involved in luteal regression (Stocco et al. 2007; Mlyczyńska et al. 2022).

# References

Arosh JA et al (2004) Prostaglandin biosynthesis, transport, and signaling in corpus luteum: a basis for autoregulation of luteal function. Endocrinology 145(5):2551–2560

Burns PD, Graf GA, Hayes SH et al (1997) Cellular mechanisms by which oxytocin stimulates uterine PGF(2 alpha) synthesis in bovine endometrium: roles of phospholipases C and A(2). Domest Anim Endocrinol 14(3):181–191

Diaz FJ et al (2002) Regulation of progesterone and prostaglandin F2alpha production in the CL. Mol Cell Endocrinol 191:5–80

Fuchs A, Behrens O, Helmer H et al (1990) Oxytocin and vasopresin receptors in bovine endometrium and myometrium during the estrous cycle and early pregnancy. Endocrinology 127:629–636

Heap RB, Fleet R, Hamon R (1985) Prostaglandin F-2 alpha is transferred from the uterus to the ovary in the sheep by lymphatic and blood vascular pathways. J Reprod Fertil 74(2):645–656

Hojo T, Skarzynski DJ, Okuda K (2022) Apoptosis, autophagic cell death, and necroptosis: different types of programmed cell death in bovine corpus luteum regression. J Reprod Dev 68(6):355–360

Jonczyk AW, Piotrowska-Tomala KK, Skarzynski DJ (2019) Effects of prostaglandin $F_{2\alpha}$ ($PGF_{2\alpha}$) on cell-death pathways in the bovine corpus luteum (CL). BMC Vet Res 15(1):416

McCracken JA, Custer EE, Eldering JA et al (1996) The central oxytocin pulse generator: a pacemaker for the ovarian cycle. Acta Neurobiol Exp 56:819–832

McCracken JA, Custer EE, Lamsa JC (1999) Luteolysis: a neuroendocrine-mediated event. Physiol Rev 79(2):263–323

Meidan R, Milvae RA, Weiss S et al (1999) Intraovarian regulation of luteolysis. J Reprod Fertil 117:217–228

Mlyczyńska E, Kiezun M, Kurowska P et al (2022) New aspects of corpus luteum regulation in physiological and pathological conditions: involvement of adipokines and neuropeptides. Cells 11(6):957

Niswender GD, Juengel JL, Silva PJ et al (2000) Mechanisms controlling the function and life span of the corpus luteum. Physiol Rev 80:1–29

Oliveira ML, Mello B, Gonella AM et al (2022) Unravelling the role of 17β-estradiol on advancing uterine luteolytic cascade in cattle. Domest Anim Endocrinol 78:106653

Penny LA, Armostrong DG, Baxter G et al (1998) Expression of monocyte chemoattractant protein-1 in the bovine corpus luteum around the time of natural luteolysis. Biol Reprod 59(6):1464–1469

Shaw DW, Britt JH (1995) Concentrations of tumor necrosis factor alpha and progesterone within the bovine corpus luteum sampled by continuous-flow microdialysis during luteolysis in vivo. Biol Reprod 53(4):847–854

Shirasuna K (2010) Nitric oxide and luteal blood flow in the luteolytic cascade in the cow. J Reprod Dev 56(1):9–14

Shirasuna K, Asaoka H, Acosta TJ et al (2004) Real-time relationships in intraluteal release among prostaglandin F2α, endothelin-1, and angiotensin II during spontaneous luteolysis in the cow. Biol Reprod 71:1706–1711

Silvia WJ, Lewis GS, McCracken JA et al (1991) Hormonal regulation of uterine secretion of prostaglandin $F_{2\alpha}$ during luteolysis in ruminants. Biol Reprod 45:655–663

Skarzynski DJ, Ferreira-Dias G, Okuda K (2008) Regulation of luteal function and corpus luteum regression in cows: hormonal control, immune mechanisms and intercellular communication. Reprod Domest Anim 43(Suppl 2):57–65

Stocco C, Telleria C, Gibori G (2007) The molecular control of corpus luteum formation, function, and regression. Endocr Rev 28(1):117–149

# Chapter 9
# Uterine Histology and Physiology: General Aspects

**Abstract** The uterine histophysiology is presented under the perspective of noting endocrine determinants of tissue modifications and interactions with defense cells. Special general consideration is given to glandular development and blood supply. Most processes are illustrated by pictures.

**Keywords** Uterus · Uterine tissues · Glands · Blood supply · Progesterone · 17β estradiol · Cortisol · Progesterone receptors · Estradiol receptors

The inner surface of the uterus is covered by a simple cylindrical epithelium. It can be changed in some parts, according to hormonal dominances to a pseudostratified epithelium with cilia. Estrogens can promote the epithelium growth, but in more degree, progesterone (P; Martin et al. 1970).

Supporting the epithelium, there is a loose connective tissue with fibroblasts, fibrocytes, macrophages, lymphatic, autonomic nerve endings, blood vessels, and, according to the phase of the estrous cycle, it is feasible to encounter cells related to immunological processes such as the presence of seminal proteins or the presence of a conceptus.

In ruminants, two different zones are seen macro- and microscopically: caruncles and intercaruncular ones. The latter has glands for embryo nutrition, and the caruncular zones are deprived of glands, but they have more blood supply than the intercaruncular zones. These zones are similar to other mammalian uteri (Figs. 9.1 and 9.2). During gestation, the caruncular zones, together with fetal cotyledons, construct placentomes where respiration takes place.

Glandular acini near the outer muscle layer are smaller than those in the subepithelium. During the luteal phase, glands develop (Díaz et al. 1986) by P action in preparation for a possible gestation, since the embryo depends on P for its survival (Mann and Lamming 2001; Spencer et al. 2004; Wang et al. 2007). Glandular atrophy was considered a cause of embryonic death (Ohtani and Okuda 1995).

The creole genotype Romosinuano (*Bos taurus*) adapted to low lands in the Colombian tropics, with a 94% fertility rate, had a greater glandular development

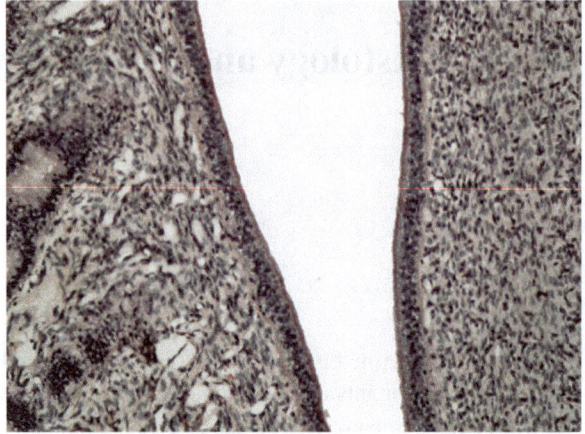

Fig. 9.1 Two adjacent
zones in the uterus of a
cow. The caruncular zone
is devoid of glands, as
opposed to the
intercaruncular one

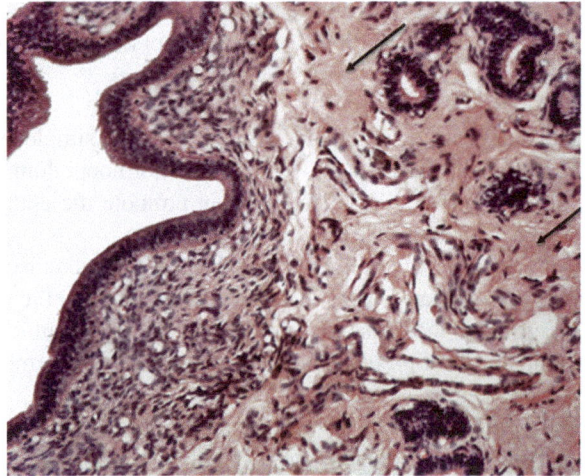

Fig 9.2 Uterus, ewe.
Estrogenic phase.
Intercaruncular (glandular)
zone. Estrogens augment
the blood flow and edema
can be seen in the lamina
propria (arrows)

than Holstein × Brahman cows, in a study carried out in Colombian low lands (Gonella et al. 2018).

Outside the connective tissue containing the glands, known as the lamina propria, there are two smooth muscle layers: a circular internal layer and a longitudinal external layer. The outermost covering is the peritoneum, composed of some connective tissue fibers and the celomic simple flat epithelium.

The capillary density in the uterus has greater values in the central parts of the uterine horn, as compared to other zones (Clavijo and Hernández 1982). The same values were encountered for the VEGF distribution (Rivas et al. 2007). In single pregnancies, the abovementioned zones are preferred for embryonic nidation in 70% of instances, in sheep (Gaviria and Hernández 1994) and cattle (Escobar and

Hernández 1996). During gestation, the sub-epithelium exhibits a development of capillary density, probably to enhance respiratory efficiency (Umaña and Hernández 1994; Sánchez et al. 2001).

Estrogens, P, and cortisol augment blood supply in the reproductive tract, especially in the uterus (Chang and Zhang 2008). This is accompanied with migration of defense blood cells to the connective tissue, as already mentioned. Also, with the increment in blood flow, edema might be present, particularly in the estrogenic phase (Fig. 9.2).

Estrogens enhance mucoid secretion in the reproductive tract, to contribute to spermatozoa ascend and removal of noxious agents, and also, in harmony with the autonomic nervous system PGF2α and oxytocin stimulate motility, in diverse reproductive circumstances, as in parturition.

The opposite actions are exerted by P. During gestation, a plug of dense mucous secretion separates the embryo/fetus from the exterior; mucous secretion and motility are minimal.

# References

Chang K, Zhang L (2008) Review article: steroid hormones and uterine vascular adaptation to pregnancy. Reprod Sci 15(4):336–348

Clavijo E, Hernández A (1982) Diferencias en la vascularización de diferentes zonas del endometrio bovino. Rev Colomb Cienc Pecu 4(1–2):39–49

Díaz F, Hernández A, Gil A (1986) Morfología endometrial y niveles de progesterona en el tejido uterino durante el ciclo estral de vacas cebú. Rev Med Vet Zoot 39:15–27

Escobar F, Hernández A (1996) Vascularización, crecimiento alantocoriónico y ubicación del embrión (o feto) durante la implantación en la vaca. Rev Med Vet Zoot 44(1):7–11

Gaviria MT, Hernández A (1994) Morphometry of implantation in the sheep. I. Trophoblast attachment, modification of the uterine lining, conceptus size and embryo location. Theriogenology 41:1139–1149

Gonella A, Ojeda OA, Lombana HG et al (2018) Serum concentration of sex-steroids, endometrial expression of their receptors, and endometrial morphology during the estrous cycle in Bos taurus Criollo and crossbred cows. J Appl Anim Res 46(1):1403–1411

Mann GE, Lamming GE (2001) Relationship between maternal endocrine environment, early embryo development and inhibition of the luteolytic mechanism in cows. Reproduction 121(1):175–180

Martin L, Finn CA, Carter J (1970) Effects of progesterone and oestradiol 17β on the luminal epithelium of the mouse uterus. J Reprod Fertil 21:461–469

Ohtani S, Okuda K (1995) Histological observations of the endometrium in repeat breeder cows. J Vet Med Sci 57(2):283–286

Rivas PC, Rodríguez-Márquez JM, Hernández A (2007) Cuantificación de células endoteliales que expresan VEGF, iNOS y eNOS en el endometrio ovino a los 20, 28 y 35 días de la gestación. Rev Colomb Cienc Pecu 20(3):280–287

Sánchez JA, Rodríguez-Márquez JM, Hernández A (2001) Área capilar sub-epitelial en el endometrio ovino en los días 0 y 14 del ciclo estral y en los 14, 20 y 24 días de gestación. Rev Cient FCV-LUZ 11(1):69–74

Spencer TE, Johnson GA, Bazer FW et al (2004) Implantation mechanisms: insights from the sheep. Reproduction 128(6):657–668

Umaña J, Hernández A (1994) Densidad capilar en el útero bovino durante la implantación. Rev
     ACOVEZ 19:10–12
Wang CK, Robinson RS, Flint APF et al (2007) Quantitative analysis of changes in endometrial
     gland morphology during the bovine oestrous cycle and their association with progesterone
     levels. Reproduction 134(2):365–371

# Chapter 10
# First Stages of Embryonic Development, Histogenesis of the Placenta, and Pregnancy Maintenance

**Abstract** In this chapter, the early embryological events are considered, together with genetical and endocrinological influences. The histogenesis of the multiplex placenta is described, as related to sheep and cow, as a gradual, slow, and fragile process, in the frame of the most relevant information on genetical molecular enhancers and moderators.

The function of the corpus luteum and some steroids, together with nutritional and immunological aspects of embryonic survival, is presented. Most processes are illustrated.

**Keywords** Embryo · Sheep · Cow · Implantation · Progesterone · Hormones · Interferon tau · Genes · Leukotrienes · Interleukins · Necrotic extremities · Embryo survival

Selection of most apt individuals for surviving is a biological strategy in evolution. This is repeated in the embryo/fetus during its intrauterine development.

During the embryonic period, the new individual is highly vulnerable; the organ systems are established. In this phase, the ruminant placenta is not completely functional, as in sheep and cow.

Embryonic survival is a biological odyssey, which entails participation of an extraordinary number of genetic and epigenetic expressions. As a consequence, multiple molecules interact, some of them known but presumably many more unveiled yet. They are produced, some in the conceptus (the embryo and associated membranes), and the remaining one in the uterus, and should function in marvelous and complex synchrony as to succeed in the production of a healthy newborn. The mother should have biological sufficiency as to maintain and accept the conceptus.

The above chapter is a translated version of chapters titled *"Primeros Estados de Desarrollo y Formación de Las Membranas Extraembrionarias, Mantenimiento de la función luteal, Progesterona y estrógenos, Implantación y placentacion, Algunos aspectos de la nutrición y metabolismo del embrión, Desarrollo vascular and Aspectos inmunológicos"* from Spanish language edition: *"Supervivencia del embrión bovino"* by Aureliano Hernández, © Editorial Universidad Nacional de Colombia 2018. Published by Editorial Universidad Nacional de Colombia.

© The Author(s), under exclusive license to Springer Nature Switzerland AG 2024
A. Hernández, *Bovine Maternal Support and Embryo Survival*,
https://doi.org/10.1007/978-3-031-62391-2_10

In sheep, sows, and cows there are huge economic losses due to embryonic mortality (EM) even when the animals are under optimal conditions of wellbeing, according to most EM reports. Around 40% of pregnancies are lost; EM accounts for 70–80% of them, and occurs between days 8 and 16 post-insemination (Peters 1996; Dunne et al. 2000; Humblot 2001; Diskin and Morris 2008).

In recent years, many research groups have identified participating genes in the expression of key molecules for embryo survival. Some of their results will be discussed, in a general manner. Nevertheless, at present, the complete map of molecular interaction, which controls embryonic development and placentation, is far for being complete.

Embryo's life can be affected by genetic and epigenetic factors and by manipulations, although it is clear that the biotechnological procedures employed have been beneficial to ameliorate reproductive performance in human beings and to conserve wildlife species in danger of extinction.

A general vision of more studied aspects related to conceptus viability is presently offered. Emphasis is placed in preimplantation and implantation phenomena, including placenta formation and an account of noxious agents, which provoke EM.

## Early Embryonic Events

The oocyte is covered by the zona pellucida (ZP), which surrounds the plasma membrane of the oocyte. After fertilization, the two fist blastomeres are surrounded by the ZP, which remain and are preserved up to the 8 days of gestation (Fig. 10.1).

The interchange of genetic information contained in the chromosomes of each pronucleus marks the genetic identity of the future embryo, as contained in the nucleus of the first cell so formed (zygote), although the mitochondrial DNA from the mother will also contribute. Epigenetic influences on the genome should also model the characteristics of the embryo/fetus and adult. In the end, live organisms are the result of genetic and environmental influences (including epigenesis).

The zygote initiates a series of mitotic divisions (cleavage ones), and the first two cells (blastomeres) are formed (Fig. 10.1).

Then, the embryo will have 2, 4, 8, 16, etc., blastomeres, which are kept together in a compact structure known as morula (a berry fruit like structure), covered by the ZP (see Chap. 1).

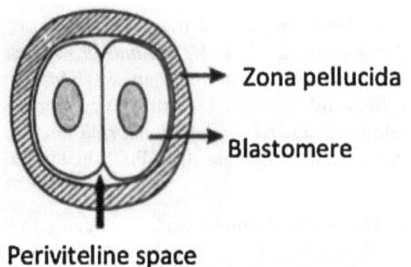

**Fig. 10.1** Schematic representation of a two-cell embryo. (Modified from Hernández and Rodríguez (2008))

Zona pellucida

Blastomere

Periviteline space

The embryo migrates being in the morula stage along the internal part of the oviduct and reaches the uterine cavity at 4 days post-fertilization (Lonergan et al. 2016).

The complete genome is in the nucleus of each blastomere. The ooplasm contains all the information needed for the first cleavage divisions to succeed. Afterward, the ooplasm will activate liberation of transcriptional factors and other signaling molecules for the building of key molecules in embryonic development. Blastomeres are totipotent, and this characteristic remains for several days. When some genes are silenced and other ones are programmed in a specific way, cells are capable of originating specific types, and they are now known as pluripotential. However, the complete genome is conserved within the blastomeres as well as in the somatic cells, although lymphocytes are an exception in this regard because, in them, there are some rearrangements. With inhibition of genetic expression, the chromatin of some regions is condensed and transcription repressed. This happens through DNA methylation. Transcription can be facilitated by histone acetylation, because chromatin gets a more open conformation. Another form of repression is carried out by the regulation of inactivated genes ("polycomb") or activated ones ("trithorax"; Veljsted 2010).

There some patterns of development appearing by virtue of signaling molecules liberation. These are mainly grouped in four families: (1) the fibroblastic growth factor family with more than 20 related proteins. (2) The Hedgehog family, which includes proteins codified by Sonic hedgehog, Desert Hedgehog, and Hedgehog Indian genes. (3) Wingless family (Wnt) with at least 15 members, which interact with transmembrane receptors known as Frizzled proteins. (4) The transforming growth factor superfamily with at least 30 molecules such as activin, the morphogenetic bone protein, the glial tissue neurotrophic-derived factor, inhibin, the nodal protein, and the anti-Müllerian hormone (Veljsted 2010).

Due to genetic restrictions, the embryo gradually losses its totipotent features. Totipotent cells lack methyl groups. This property is present in the internal cell mass components; they lack methyl groups in the DNA of specific-type cell's genes, although there are some unusual methylation patterns in them (Gilbert 2003).

Based on genetic studies, there is an increasing understanding of blastomere differentiation in internal cell mass or trophoblast cells. Presence or absence of transcription factors defines the blastomere's destiny. For instance, in mice, at the eight cells state of development, it is already clear which cells will occupy an internal or external position in the embryo (Johnson and Ziomek 1981).

In the bovine, the mechanisms of determination and destiny of zygote-derived cell lines are similar to those found in mice, although the definitive differentiation of the trophoblast happens at a different stage of pregnancy as compared to that process in mice and hemochorial placentae. In fact, in the latter, implantation takes place within the first week of pregnancy, and in the cow, it begins at around day 19 (Dlaikan et al. 1999; Pfeffer and Pearton 2012), and it is complete during the third month of gestation, taking into account that the placentomes (the fusion of uterine caruncles with fetal cotyledons) are gradually established.

The mouse, cow, and human embryos, at the two cells stage, have a self-produced preimplantation factor (PIF), which would induce maternal recognition in later stages of pregnancy (Roussev 1996; Barnea 2004; Stamatkin et al. 2011). The PIF identified by the Elisa test, would then be an in vitro marker of embryo viability, absent in nonviable mouse and cow's embryos. This molecule might act in the embryo's preparation for immunological recognition and to develop as a blastocyst (Stamatkin et al. 2011).

Two more substances, the early pregnancy factor (EPF) and platelets activating factor (PAF) secreted by human and mouse embryos (Minhas et al. 1996), are other potential viability markers. The EPF present in platelets (Cavanagh and Morton 1994) has immunosuppressive properties (Noonan et al. 1979) and could be important in embryonic survival, as a possible cellular proliferation enhancer.

The microenvironment surrounding the embryo is a key factor for its development. According to their transcriptomic studies, Salilew-Wondim et al. (2013) proposed that the genetically determined embryonic, oviductal, and uterine biochemical composition is crucial for the success or failure of gestation.


## The Blastocyst


At the end of the morula stage, some cells separate and become localized at one pole of the embryo. They form a compact structure known as the internal cell mass (ICM). The remaining cells constitute a one-layered coat placed at the internal surface of the zona pellucida, the hypoblast, which originates from the inferior part of the ICM (Fig. 10.2). The dorsal area of the ICM is known as the epiblast (Figs. 10.2 and 10.3). Most part of the embryonic body is formed from this zone. The appearance of the ICM marks the establishment of the dorsoventral axis of the body.

At the seventh day of gestation, the blastocyst has an external layer, the ectoderm (trophoblast or trophoectoderm), which forms a continuum with the Rauber's layer, a short layer of cells located above the ICM that replaces the trophoblast cells in this area (Fig. 10.3).

The zona pellucida is lost around day 8 of pregnancy, and the blastocyst is then exposed to the intrauterine environment, surrounded by epithelial secretions. The rupture of the zona pellucida, at least in the mouse embryo, is carried out by the action of the enzyme strypsin secreted by the parietal trophoblast (Perona and Wassarman 1986; O'Sullivan et al. 2001).

At the end of the blastula stage, it is not feasible to obtain genetically identical individuals from single cells of the embryo, as previously, in the morula stage. From

**Fig. 10.2** Illustration of a 6-day-old bovine blastocyst, before eclosion. The primitive vitelline space is also known as blastocele

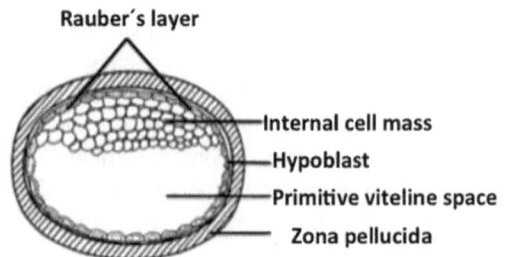

Rauber's layer

Internal cell mass

Hypoblast

Primitive viteline space

Zona pellucida

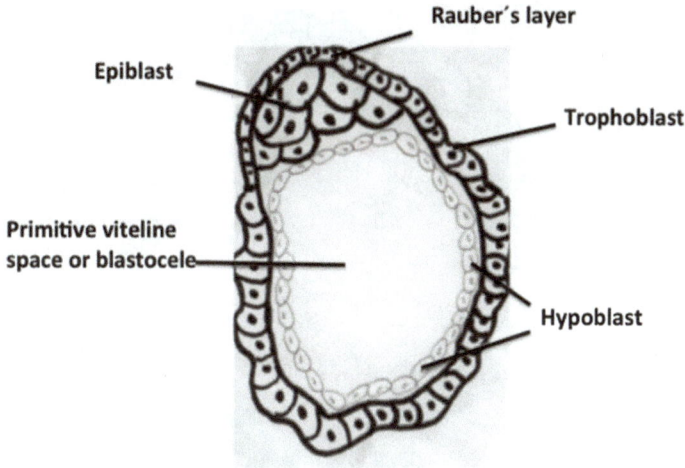

**Fig. 10.3** Blastocyst, liberated from the zona pellucida. More than 8 days old

**Fig. 10.4** Scheme of part of an embryo to show the origin of the primitive streak from the caudal region of the embryonic disk. Mesodermal cells arise from the primitive streak

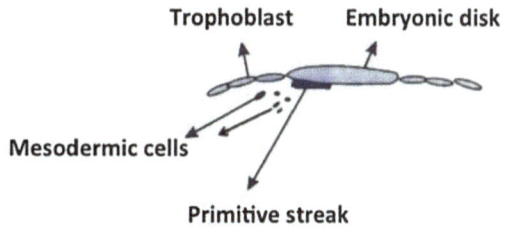

there on, the specialization of cells ensues as to gradually build up the four basic tissues of the body.

## The Extraembryonic Membranes

From the epiblast, in the future caudal zone of the embryo, an accumulated mass of cells can be seen as a thickening of the embryonic disk, known as the primitive streak (PS; Fig. 10.4). From there, the first endodermal and mesodermal cells will appear. The PS marks the craneo-caudal embryonic axis.

The mesodermal cells expand in all directions between the ectoderm and the endoderm. At each side of the longitudinal axis, the mesoderm engrosses. This thickened area is named the dorsal or paraxial mesoderm. It continues laterally as the intermediate mesoderm, which outgrowths at each side as the lateral mesoderm. The latter is composed by an external layer, the somatic mesoderm, which underlines the trophoblast, thus forming the chorion or somatopleure (bilayered), and an internal layer, the visceral or splanchnic mesoderm, which surrounds the tubular

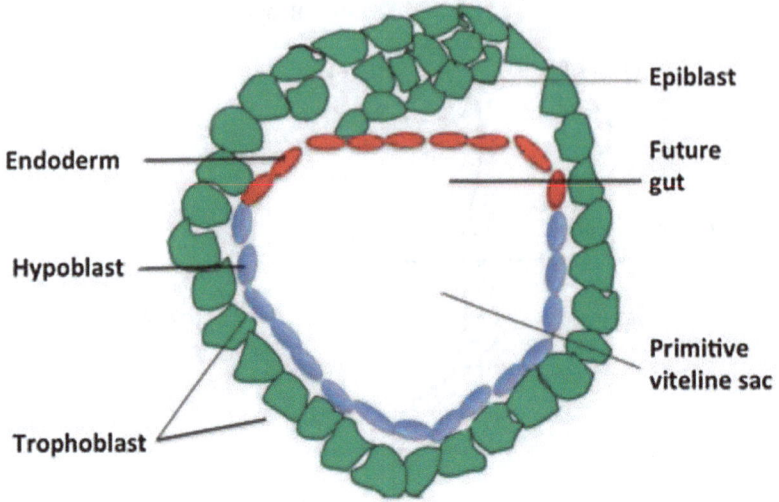

Fig. 10.5 The zona pellucida and the Rauber's layer have disappeared and the endoderm is present

endoderm, thus forming the splachnopleure that constitutes the wall of the intestine (Figs. 10.5, 10.6, 10.7, 10.8, and 10.9).

At first, the endoderm is formed by a single cell layer, which occupies the adjacent region of the ventral part of the ICM. In the lower region, the endoderm integrates a single chain of cells with those of the hypoblast (Veljsted 2010; Figs. 10.5, 10.7, and 10.8). Later on, as already explained, the endoderm will form, with the visceral mesoderm, the intestine. The hypoblast gives rise to the internal wall of the yolk sac (see Gilbert 2003; Figs. 10.5, 10.7, and 10.8). Later on, there will be a communicating channel between the mid-intestine and the yolk sac through the yolk stalk (Fig. 10.8). The yolk sac regresses a few days after its appearance.

The cells of the epiblast migrate toward the primitive streak under the control of the fibroblastic growth factor 8 (FGF8), which inhibits cadherin, a molecule responsible of the cells union in the epiblast. FGF8 controls mesodermic cell specialization through the expression of Brachyury gen (Sadler 2006). Likewise, by expression of the gen goosecoid. Each cell has around 500 activin receptors. Activin is necessary for specification of cells in the mesoderm. If there are 100 receptors bound, the Brachyury gen is expressed and the cells will form the ventral mesoderm and the lateral mesoderm. When the binding involves 300 receptors, the cells will express the goosecoid gen and will be differentiated to constitute the dorsal mesoderm (Gilbert 2003).

In the most anterior part of the primitive streak, a condensation of cells is formed, known as the Node (or Node of Hensen). The notochord, which derives from the Node, is a central chord located along the longitudinal axis of the embryo (Figs. 10.7 and 10.8), the antecessor of the spinal column in fish. In mammals, the notochord remains as the pulpous chore of the fibrocartilage intervertebral disks (Patten 1948). It is generally claimed that the notochord constitutes an ancestral reminiscence, a

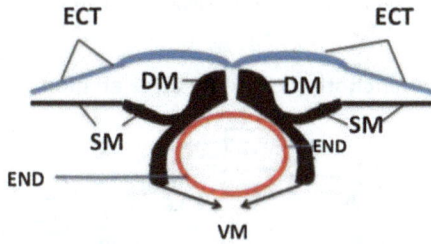

**Fig. 10.6** Schematic representation of a transversal section of an embryo as to show the mesoderm divisions: dorsal mesoderm (DM), which at each side becomes intermediate mesoderm (not indicated). This divides laterally as a twofold layer known as the lateral mesoderm, which splits in the somatic mesoderm (SM), accompanying the ectoderm or trophoblast (ECT) and the visceral or splanchnic mesoderm (VM), surrounding the endoderm of the future intestine. The arrows indicate the direction of movement of the VM

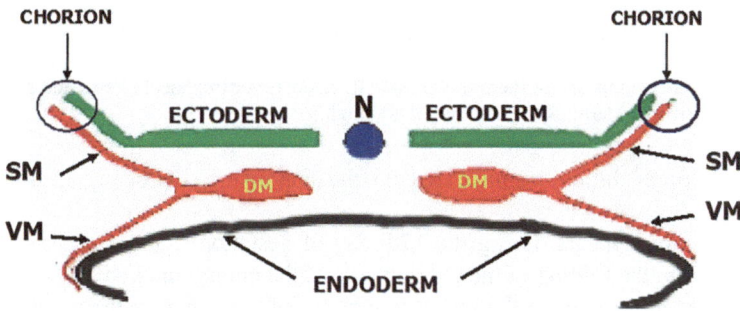

**Fig. 10.7** Schematic representation of the three primitive layers during the amnion formation. The chorionic ends will become the amniotic folds. These will meet above the ectoderm in the middle zone, as to form the amnion. SM: somatic mesoderm. DM: dorsal mesoderm. VM: visceral (splanchnic mesoderm)

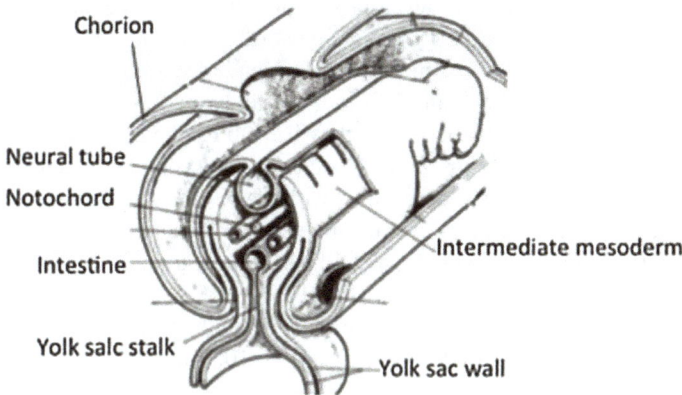

**Fig. 10.8** The amniotic folds (AF) moved downward and are deepened, thus forming the future amniotic cavity (3). From the middle part of the intestine (2), an invagination develops as the yolk sac (YS, cut). At the dorsal zone of the future embryonic body, the ectoderm (ECT) and the somatic mesoderm (SM) fuse in the middle part and then continues as the chorion. 1: neural tube. According to Marrable, WA, 1971 by permission

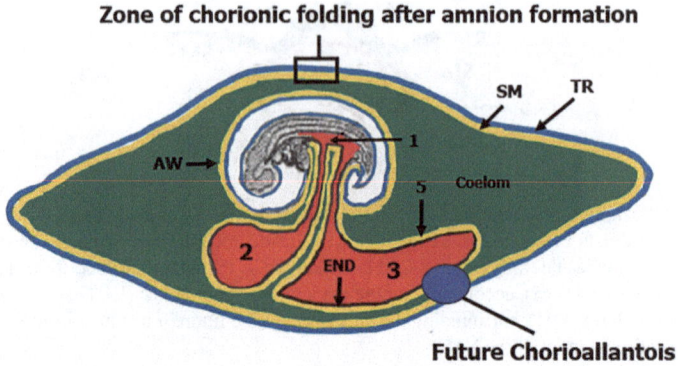

**Fig. 10.9** The yolk sac (2) emerges from the midgut (1) and the allantois (3) from the caudal one. Both of these invaginations have an external coat composed by visceral mesoderm (5) and an inner one of endoderm (END). The allantochorion is formed by the union of the allantois with the chorion. The latter is composed of the trophoblast (TR, ectoderm or trophoectoderm) and an internal coat of somatic mesoderm (SM). (Patten (1948), modified)

testimony of evolution, summarized on the aphorism: "Ontogeny recapitulates phylogeny."[1]

The amnion appears during the 17th day of gestation (Sandra et al. 2015). It develops from the folding of the chorion around the embryonic body. The amniotic folds meet and fuse above the embryonic body, and the chorion then continues to grow beyond the limits of the embryo (Figs. 10.7, 10.8, and 10.9). Furthermore, the amniotic folds approximate laterally to intrude the embryonic body in the amnion cavity. The intestine is then included in the embryonic body, and finally, two ventral openings (invaginations) of the intestine can be seen in the mid and caudal areas of the embryo: the yolk sac and the allantois, respectively (Figs. 10.8 and 10.9). The yolk sac regresses soon after its development, and the allantois expands as to reach the chorion until it does so at 25 and 28 days of gestation in the ewe and cow, respectively (Hernández 1971; Jiménez and Hernández 1982).

The main function of the allantois is to make up blood vessels by the processes of vasculogenesis and angiogenesis, which will establish the interchange of respiratory gases with the maternal circulation, as the placenta is being structured.

The chorionic somatic mesoderm fuses with the visceral mesoderm of the allantois and thus becomes a unique layer of mesenchyme beneath the trophoblast. The allantoic endoderm is located in the internal part of the mesenchyme and surrounds the external part of the allantoic cavity (Fig. 10.9).

It should be noted that blood vessels are formed in the visceral mesoderm, when the mesoderm fusion has not taken place yet (Fig. 10.10).

---

[1] The recapitulation theory advanced by John Hunter (1728–1793) and Carl Friedrich Kielmeyer (1975–1844) was disseminated by Ernst Haeckel (1834–1919).

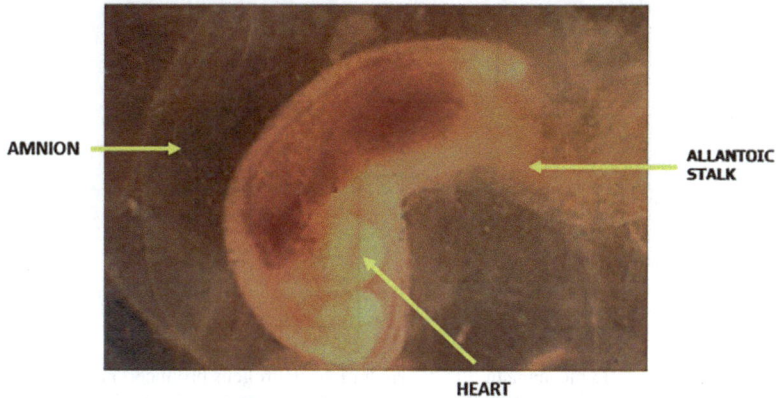

AMNION →                                                                    ALLANTOIC
                                                                            STALK ←

                                    HEART

**Fig. 10.10** Bovine embryo, 23 days. The allantois emerges from the caudal part of the embryo and the amnion surrounds it

## Luteal Function Maintenance

During the day 16th of pregnancy, interferon tau (IFNT) produced in the trophoblast (TR), indirectly acts on the corpus luteum to maintain and increase progesterone (P) secretion, the main known hormonal support of pregnancy. Before day 16, the serum levels of P are similar to those found during the estrous cycle (See Fig. 5.1; Forde et al. 2011).

IFNT was known as trophoblastic protein 1. It represents one of the five families of type 1 interferons, and it was derived from the omega interferon 36 million years ago, and in bovines, 10–15 million years ago. The ruminant's interferon gene retains the ancestral characteristic of lacking introns, although the reason for this is unknown. Both IFNT and interferon omega contain 172 amino acids. IFNT has proliferative properties, but it is mostly known because it functions as an antiluteolytic agent (Demmers et al. 2001).

At the present time, several IFNT characteristics in bovine, ovine, caprine, and some other ruminants are recognized. IFNT is elaborated in the RSDA2 domain (Radical S-adenosyl methionine domain containing). A greater quantity of mRNA IFNT expression is found at days 17 and 19 of gestation, than at days 14 and 25 (Ealy et al. 2001). In the cow, trophoblastic IFNT is not detectable in the 25th day of gestation (Demmers et al. 2001).

The ovine IFNT induced by IFIH1 (helicase C domain 1) is produced in the TR during preimplantation and implantation. In turn, IFNT provokes IFIH1 and RSDA2 expression (Song et al. 2007).

IFNT acts on the endometrium to induce maintenance of luteal function. It also promotes uterine receptivity to TR adhesion. Once contact between the two structures occurs, IFNT expression ceases (Demmers et al. 2001). Its liberation from the trophoblast aims to eliminate pulsatile PGF2α liberation, thus inhibiting luteolysis. IFNT binds its epithelial uterine receptors. Later on, the expression of estrogen

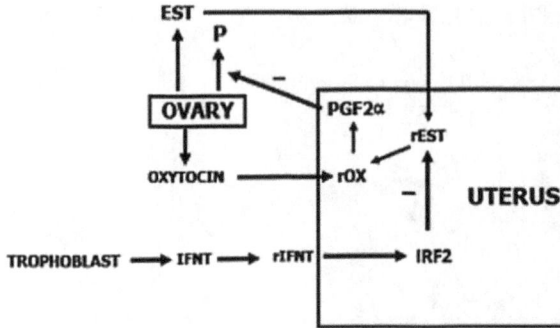

**Fig. 10.11** Antiluteolytic mechanism. Prostaglandin F2α. Estrogens produced in the ovary binds with uterine receptors. This complex stimulates oxytocin attachment to its uterine receptors as to provoke PGF2α secretion, which, as the main luteolytic molecule, reduces progesterone secretion. The trophoblast secretes interferon tau (IFNT). The latter binds its receptors (rFNT) to stimulate the inhibitory IRF2 to block the positive effector of estrogen on PGF2α action

alpha receptors is suppressed by IRF2 action, which hampers the transcription of estrogen receptors. In absence of the latter, estrogen cannot induce expression of oxytocin receptors, a must for uterine PGF2α luteolytic pulses of secretion; hence, the corpus luteum continues P production and liberation (Spencer et al. 1996; Bazer et al. 1996, 2012b). In Fig. 10.11, there is a synthesis of the antilueolytic mechanism.

IFNT inhibits the movement of PGF2α (Banu et al. 2010), which might then be a complementary mechanism in luteolysis.

There are 12 cDNA for IFNT in the conceptus. It is not known whether or not this cytokine represents allelic forms of the same gene. Some IFNT variants are localized in chromosome 8, which is associated with IFN gamma and omega (Walker et al. 2009). IFNT regulates various genes in the endometrium, related with growth, angiogenesis, and extracellular matrix deposition (Chen et al. 2006). The expression of IFNT variants was similar in male and female blastocysts, and although the expression of some of them differed with age, predominant variants in male and female blastocysts were boIFN1a and boIFN3c (Walker et al. 2010).

Eomesodermine T box is a protein present in the trophoblast during the 22nd day of gestation. It was suggested as a regulator of IFNT transcription (Sakurai et al. 2013a). In a similar study, two types of IFNT genes were found in the cow's uterus, although the route of transcription is not the same (Sakurai et al. 2013b).

The fibroblastic growth factors (FGF) 2 and 10, elaborated in the ewe's lining uterine epithelium, induce cellular migration, which is a necessary event for implantation to succeed. The FGF 10 could have an influence in IFNT release, since its production is elevated around day 16th of gestation (Cooke et al. 2009; Yang et al. 2011). P stimulates FGF 10 production (Satterfield et al. 2006), and synthesis of various types of FGF increases from the morula stage up to the 28th day of pregnancy.

FGF uterine expression in the cow is associated with growth of the trophoblast (Ozawa et al. 2013) and expression of receptors related to different FGF changes during days 7–19 of gestation, in spite that there is not a specifically role as related to preimplantation events (Okumu et al. 2014). In the same context, evidence was given for the FGF2 stimulation of INFT build up (Michael et al. 2006). Information is lacking about the role of FGF's during implantation, particularly as related to P and IFNT functions.

The granulocyte-macrophage stimulating colony forming factor (GM-SCF) induces IFNT production in ewes (Imakawa et al. 1997).

In studies carried out in mice, lack of GM-SCF in the mother or the embryos was a cause of prenatal death. If both, the mother and the embryos lack GM-SCF, and the problem worsens (Robertson et al. 1999). Some granulocytes elaborate GM-SCF under ovarian steroids, embryonic signals, and seminal fluid control. Implantation and granulocyte function are correlated (Reviewed by Robertson 2007).

Pennington and Ealy (2012) provided experimental evidence about the role of the morphogenetic proteins 2 and 4 in IFNT synthesis.

In spite of advances in regulation of IFNT secretion, full knowledge in this context is incomplete (Ealy and Yang 2009).

IFNT activation of protein secretion appears as essential for pregnancy success. Within them, one ubiquitin-like protein (Rueda et al. 1993; Staggs et al. 1998) and the chemotactic granulocytic protein 2 (Texeira et al. 1997) were found. The latter could intervene in early events in gestation, such as angiogenesis and trophoblast adhesion to the uterine lining epithelium (Staggs et al. 1998).

An overview of participating molecules in IFNT production is given in Fig. 10.12.

According to Demmers et al. (2001), in clinical cases of luteal insufficiency, it would be more beneficial to use P instead of IFNT. As already mentioned, at 25 days of gestation, IFNT production is not sufficient as a luteotrophic agent (Demmers et al. 2001), and it is expected that there should be an alternative mechanism

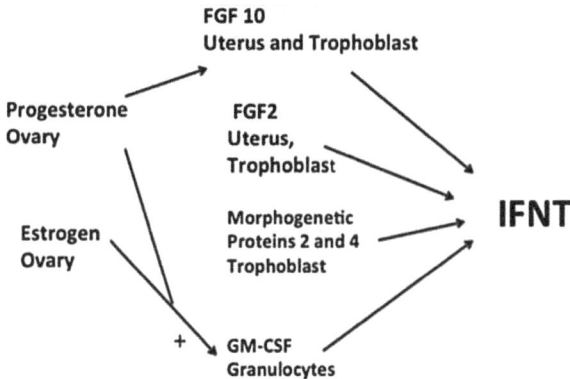

**Fig. 10.12** Molecules that stimulate IFNT secretion are produced in the ovaries, uterus, and trophoblast. Progesterone and estrogen induce granulocyte-macrophages stimulating factor (GM-CSF). FGF2: fibroblast growth factor two. (Information extracted from Imakawa et al. (1997), Robertson et al. (1999), Michael et al. (2006), Satterfield et al. (2006), Robertson (2007), Cooke et al. (2009), Yang et al. (2011), Pennington and Ealy (2012), and Ozawa et al. (2013))

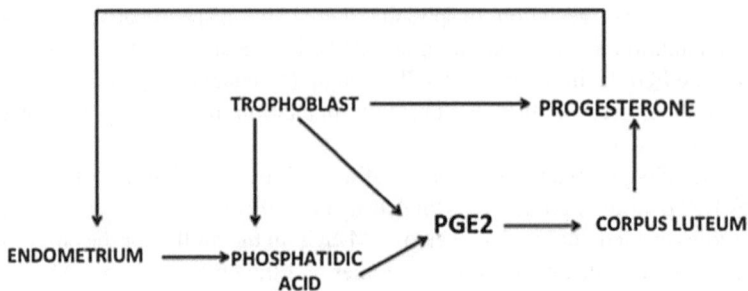

**Fig. 10.13** Progesterone production by the trophoblast (including the binucleate cells) and luteal function maintenance by prostaglandin E2 (PGE2), which could take over the interferon tau luteotrophic action. The phosphatidic acid, produced in the trophoblast and uterus, stimulates PGE2 liberation. (Information obtained in Low and Hansen (1988), Woclavek-Potocka et al. (2010), and Lisewska et al. (2009))

Prostaglandin E2 is a possible candidate (PGE-2; Weems et al. 2002); it should be taking into account that the binucleated cells of the trophoblast (BNC) produce P and PGE-2 (Low and Hansen 1988).

Phosphatidic acid is a phospholipid produced in the uterus and the trophoblast, which stimulates PGE-2 liberation (Woclavek-Potocka et al. 2010; Liszewska et al. 2009; Fig. 10.13).

## Some Steroid Activities

The most studied steroid hormones as pertinent to early gestation are P and estrogens. Their serum levels should be in adequate balance for pregnancy success (Conley and Assis-Neto 2008). Estrogens levels must be lower than those detected during the estrogenic phase of the estrous cycle.

P prepares the endometrium for embryo implantation, inhibits the generation of luteinizing hormone pulses during pregnancy, stimulates endometrial secretion to procure embryo/fetus wellbeing, and prevents muscle uterine contractions in close harmony with the autonomic nervous system. Supplementation with exogenous P increases the probabilities of embryonic survival (Mann et al. 2001).

Maurer and Echternkamp (1982) proposed that P serum levels in the initial phases of corpus luteum formation could be important for endometrial receptivity, and, hence, for embryonic survival.

As expected, the development of the corpus luteum depends on the adequate IFNT serum levels (Mann and Lamming 2001).

In recent years, it has been increasingly evident that P and IFNT act in a coordinated way to control embryo's nutrition and the growth of uterine and conceptus tissues. These processes involve the control of genes expression that are compromised in trophoblast adhesion to the lining uterine epithelium and phenotypic modification of stromal cells in the endometrium. In the same direction, P and estrogen receptors expression would be silenced as well as the expression of some antibodies. The cell membranes permeability would be also altered, and the maternal and conceptus interchange of molecules facilitated. FGF-10, the insulin growth factor (IGF) receptor, and the IGF-1 and IGF-3 are regulated by P during preimplantation (Satterfield et al. 2006).

One of the main functions of P is to induce changes in the expression of necessary molecules for embryonic nutrition, particularly before placentation. No differences in hormonal expression were found when comparing the estrous cycle with early pregnancy before days 15–16 (Forde et al. 2011). In a comparative study using cows, one group was given exogenous P4. In this group, P induced a greater nutrients elaboration in the uterine glands than in the nonsupplemented one, such as triglycerides and glucose. Also, other crucial elements for embryonic development were found in the treated group of animals as compared to the control one (Forde et al. 2009).

Endometrial secretion and glucose and amino acid synthesis were enhanced by P administration (Mullen et al. 2014). According to Gao et al. (2009), glucose transport is controlled, in the ewe, by P and IFNT between days 10 and 15 of gestation.

P is also produced in the uterine caruncles. The TR and endometrium can regulate P availability, use by the action of $20\alpha$ hydroxysteroid dehydrogenase enzyme, which metabolizes P to an inactive compound (20 alpha progesterone; reviewed by Nguyen et al. 2012).

During the first trimester of gestation, the TR is a source of estrogens. Estradiol and estrone were found in caruncles and villous trophoblast from the 4th month of pregnancy on (Tsumagari et al. 1993) and by the end of gestation, estrone becomes the major estrogen (Hoffmann et al. 1997).

In a hot and humid weather of the tropics (Montería, Colombia), the creole genotype known as Romosinuano, well adapted to climate conditions, showed the best reproductive efficiency as compared to cross-bred genotypes (believed as nonadapted Zebu × Simmental and Zebu × Holstein). The mentioned parameter was correlated with higher serum P levels in the luteal phase of the ovarian cycle (Grajales et al. 2006; Grajales and Hernández 2008).

The development of the uterine glands and P serum levels were higher in Romosinuano and "Costeño con cuernos" (another creole well-adapted breed) genotypes than in Zebu cows during the estrous cycle (Gonella et al. 2010). Likewise, they had more estrogen and P receptors in the endometrial glands (Gonella 2011).

## Elongation of the Blastocyst

The chorion of the bovine blastocyst initiates a rapid growth during the 15th day of pregnancy (Spencer et al. 2008), which could be interpreted as a strategy to augment the number of trophoblastic cells for physiological exchange with the uterus to supply the metabolic demands of the rapid developing conceptus.

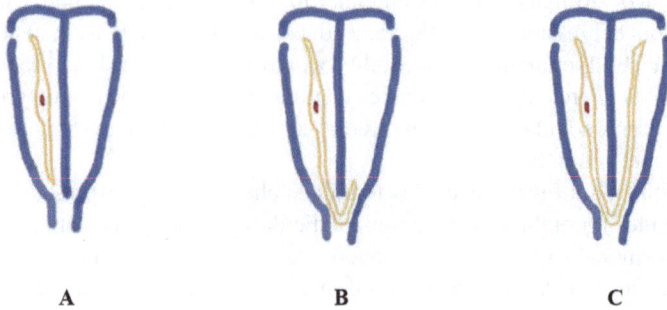

A                  B                  C

**Fig. 10.14** Drawing to show the elongation of the bovine (**a**) at 16 (**b**) 22 and (**c**) 28 days

At 14 days of gestation, the conceptus is 6 mm long, 6 cm at day 16, and 20 cm at 19 days (Betteridge 1980; Berg et al. 2010; Fig. 10.14). During the 24th day of pregnancy, it might occupy both uterine horns (Maddox-Hyttel et al. 2003). It should be noted that in cows and sheep, there are individual variations in the velocity of elongation, and hence, in the size of conceptuses (Escobar and Hernández 1996; Gaviria and Hernández 1994).

The increment in the number of TR cells due to blastocyst elongation is responsible for the increased elaboration of IFNT and not due by upregulation of IFNT mRNA (Robinson et al. 2006). Ledgard et al. (2011) found that more prolific cows had bigger conceptuses, higher P serum levels and protein content in the uterine fluids, than less prolific animals.

Elongation and survival of the conceptus are controlled by some prostaglandins and cortisol (Dorniak et al. 2011, 2012, 2013; reviewed by Brooks et al. 2014). Prostacyclin could also contribute (Cammas et al. 2006).

Elongation and growth of the conceptus are under the control of P (Garrett et al. 1988; Mann et al. 2001; Satterfield et al. 2006). The presence of a conceptus and P concentration alter the expression of genes, such as APOA1, ARSA, LCAT, NCDN, and members of the PLIN family were proposed to participate in chorion elongation (Forde et al. 2013).

Concentration of P in maternal circulation acts through the endometrium and is associated with different rates of chorionic elongation. The embryo's maintenance depends on IFNT secretion, which appears to have a similar P effect during the blastocyst stage, given that there is evidence that IFNT controls the expression of several genes and proteins in the uterus (as P does), from the blastocyst stage up to the peri-implantation period (Forde and Lonergan 2012).

Periostin, which commands osteoblasts recruitment in more advanced stages of development, promotes adhesion and dissemination during elongation, under the control of P, according to Ahn et al. (2009).

## Implantation and Placentation

Some of the conditions needed for implantation to succeed are as follows: (1) absence of immune rejection, (2) maternal wellbeing, (3) suited uterine microenvironment, and (4) appropriate genetic constitution of the embryo.

Although the rapid blastocyst elongation is generally framed between days 13 and 19 of gestation, the chorion continues its expansion, and the trophoblast should make contact with the lining uterine epithelium. This marks the beginning of implantation of the conceptus.

As explained in Chap. 9, the uterine histology changes according to the P or estrogens' influence. Estrogens augment blood flow, edema formation, and defense cells migration, as well as motility related to the development of cilia on the uterine lining epithelium and muscle contraction. P inhibits muscular movement and promotes glandular development (Fig. 10.15).

Initially, the TR is a simple flattened or cuboidal epithelium, which later on will develop into a two- to four-layered epithelium (Figs. 10.16, 10.17, and 10.18). The first anatomical modifications of the ULE, which appear as a consequence of TR attachment, are seen at day 16th of gestation in sheep (Guillomot 1995) and at day 20 in the cow (Leiser 1975; Fig. 10.16).

After TR adhesion, the TR cells proliferate and the following tissue changes are present: (1) the development of the trophoblast and the diminution in height of the ULE, from a cylindrical simple or pseudoestratified epithelium to a flattened type of tissue (Figs. 10.17, 10.18, and 10.20). Syncitia of the ULE are seen between days 20 and 23 of gestation (King et al. 1981; Fig. 10.20). The flattening of the ULE

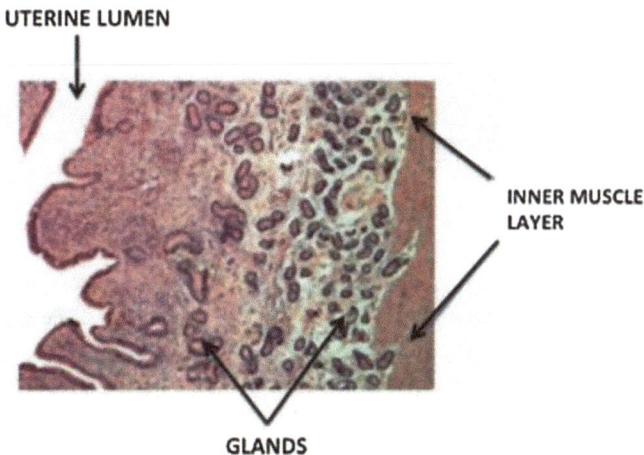

**Fig. 10.15** Uterine histology. Note the difference between the glandular development of the superficial zones

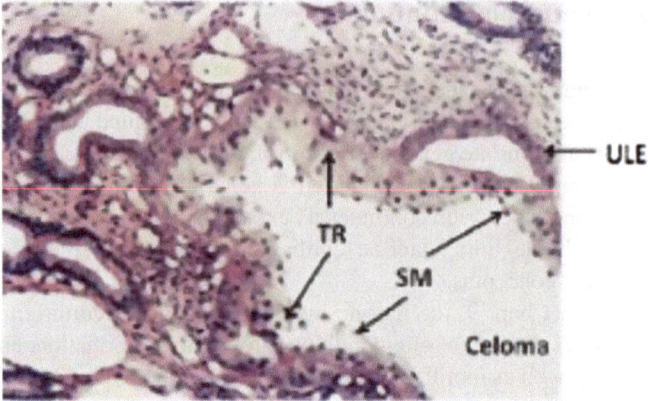

**Fig. 10.16** The uterine lining epithelium is not modified in the zones where trophoblast attachment is lacking (ULE). TR: trophoblast. SM: somatic mesoderm. Sheep, 23 days of gestation. (Hernández 1971)

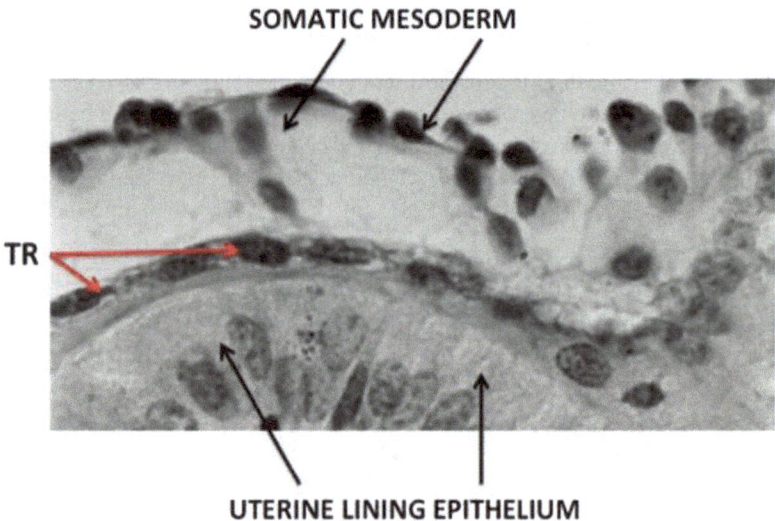

**Fig. 10.17** Initial attachment of the trophoblast (TR) to the uterine lining epithelium. The TR is still a nondeveloped tissue and the uterine lining epithelium remains intact (nonmodified)

sometime after TR adhesion has not a satisfactory explanation at present. It could be speculated that it occurs as to diminish the diffusion distance of gases between the blood vessels of the embryo/fetus and the maternal ones, later on, when the placenta functions as a developed respiratory organ. The diffusion distance in Chorioepithelial placentae of ruminants is greater than the one seen in hemochorial and endotheliochorial placentae and presumably less efficient as a respiratory organ (Bartels 1970).

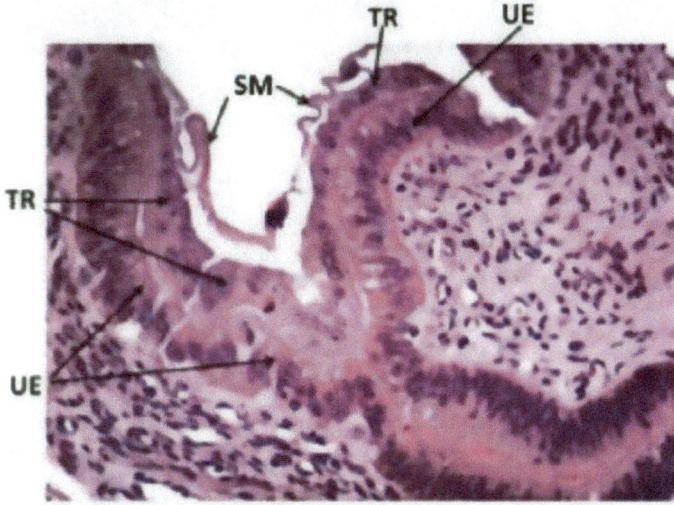

**Fig. 10.18** Developing bovine trophoblast in contact with a nonmodified uterine lining epithelium

In the developed TR, there are some binucleate cells (BNC), which can be seen from day 18th of gestation onward (Dlaikan et al. 1999), that is, at the beginning of implantation.

The histological changes of TR development do not occur simultaneously all over the chorion. In fact, whereas in the vicinity of the embryo the TR is predominantly well developed, in more distant zones, the TR is simple cuboidal. These anatomical variations are present in sheep between days 14 and 24 and in the cow between days 18 and 35. After those days the TR, it is a two- to four-layered epithelium in the major part of its area (Gaviria and Hernández 1994; Dlaikan et al. 1999).

In summary, it is possible to find the following histological characteristics of the forming placenta, during days 16–28 of pregnancy in sheep, and between days 18 and 36 in the cow: (1) nondeveloped TR: unmodified ULE (Fig. 10.17). (2) Developed TR: intact ULE (Fig. 10.18). (3) Developed TR: flattened ULE (Fig. 10.20).

The chorion is avascular and the blood vessels in its underlining connective tissue, when the allantochorion has been formed, derive from angiogenic islands. These originate from the mesenchyme of the visceral mesoderm within the allantois (Figs. 10.19 and 10.20).

The initial period of allantoic formation in sheep and bovines coincides with expression of several genes. Thus, allantoic formation was associated with the expression of the following genes: type XII collagen alpha 1, type 2 collagen alpha 2, epsilon 4 beta-globin, osteonectin, and uroplakin (Ledgard et al. 2011). As time goes on, the collagen fibers in the mesenchyme are seen in increasingly more dense packages. In this context, Shang et al. (1997) identified the pro-collagen gen III and proposed a relationship with expansion of the allantois.

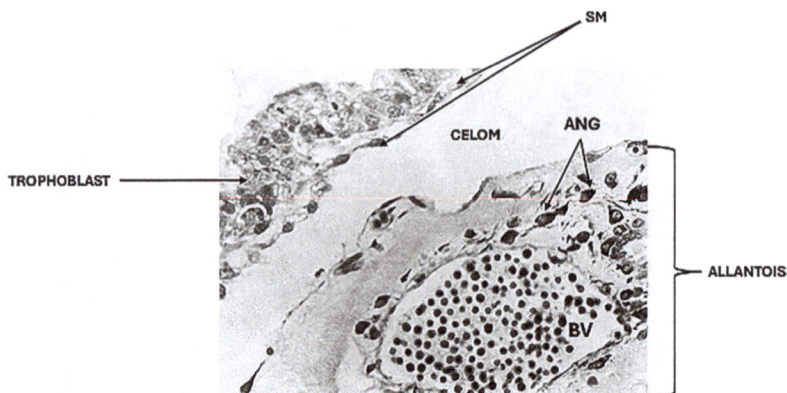

**Fig. 10.19** The trophoblast is now a stratified epithelium. The allantois is reaching the chorion and the angioblasts (ANG) are proliferating as to form blood vessels (BV). (Hernández 1971)

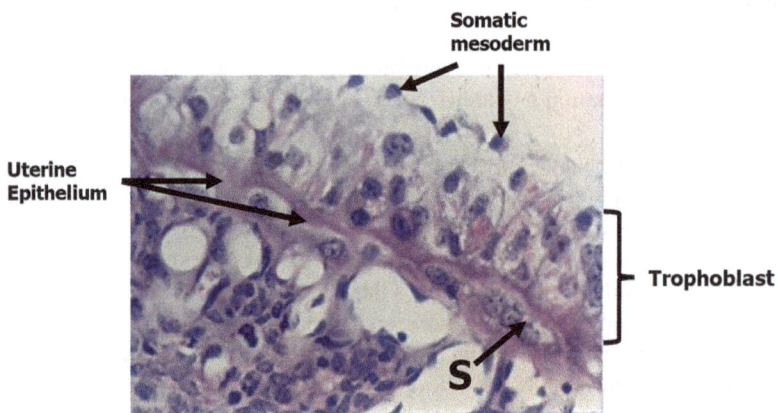

**Fig. 10.20** The uterine lining epithelium (UT EP) has become flattened. Here, it has a syncitium (S). Capillaries are empty and dilated because of the method of fixation (perfusion). The asterisks indicate the fused trophoblast and UT EP cell membranes. Some eosinophilic crystalloids are present in the trophoblast. (Hernández 1971)

In the extracellular matrix remodeling of the endometrium and the chorioallantoic mesenchyme, the ADAMTS1 protease intervenes, codified by WNT, acting as a desintegrin and metalloproteinase. It is possibly expressed in the TR and the endometrium under P induction, from preimplantation and implantation up to advanced stages of gestation (Mishra et al. 2013).

Serines and cell adhesion molecules also play a role in implantation given that TR and ULE cell membranes should fuse during the adhesive stage. A uterine secreted phosphoprotein binds to TR and ULE integrins to facilitate attachment and also the cellular interaction of the extracellular matrix and cytoskeletal remodeling (Kim et al. 2010).

Osteopontin, an extracellular matrix protein, could be a mediator in implantation in the ewe, acting as a bridge between integrins. These are expressed in TR and ULE (Johnson et al. 2001).

The carrier protein of the insulin growth factor 1 could also be compromised in migration and adhesion of TR (Simmons et al. 2009).

Galectins are glycoproteins, which bind their receptors at the cell surface. They play roles in apoptosis, chemo-attraction, and cell migration, as well as in cell growth and differentiation. They are found in ULE by P and IFNT induction during implantation in sheep; galectins form crystals in the TR, which could be the crystalloids, were described by Amoroso (1952).

Between days 30 and 120 of gestation, galectin 1 was encountered in the endometrial connective tissue and the trophoblastic binucleate cells, and galectin 3 in the ULE. Both galectins 1 and 3 were encountered in the placentomes. Galectin 4 was encountered in the ULE and binucleate cells, between days 220 and 275 of gestation (Froehlich et al. 2012).

MacIntyre et al. (2002) provided evidence that the fusion of binucleate cells with the ULE would stimulate changes in integrins and extracellular matrix in the endometrial sub-epithelium.

It is feasible that arginine, leucine, and glucose present in the secretion of the uterine glands provoke TR cells proliferation and migration (Kim et al. 2011).

Lipophosphatidic acid is secreted in the endometrium and the conceptus; it regulates PGE-2 liberation (Woclavek-Potocka et al. 2010) and the TR mitotic index (Liszewska et al. 2009).

Small quantities of the vascular adhesive molecule 1 (VCAM1) are produced in the ULE and uterine glands of the intercaruncular zones, in the stroma, and the vascular endothelium of the caruncular zones, during the 17th day of gestation. Between 20 and 22 days, the amount of VCAM1 has increased. It is likely that VCAM1 liberation is stimulated by secretions from the trophoblast and uterine glands. The trophoblastic $\alpha$ 4 integrin together with VCAM1 could be involved in adhesive processes attained to TR: ULE interaction (Bai et al. 2014).

Ghrelin, a 28 amino acid peptide, known to stimulate growth hormone secretion and food ingestion, is produced in the stomach and the hypothalamus. It was proposed to play a role in reproductive events including embryonic implantation (Luque et al. 2014).

Bombesin, a peptide related to the gastrin liberating factor (GRP), produced in the trophoblast and the uterus, could participate in the development of the ovine placenta (Beltrán 2000). In the ovine uterus, bombesin is produced in the glands at day 16 of the estrous cycle and in gestation, at days 20 and 140 of gestation (Whitley et al. 1998). Budipitoj et al. (2001) identified the GRP in the two components of the placenta of the cow and the uterus of nonpregnant cows, particularly, in the glandular epithelium. Bombesin and GRP were proposed to participate in ionic transport (Matthews et al. 1993) and modulation of myometrial contractions (Amiot et al. 1993).

The leukemia inhibitory factor (LIF) could have a function during implantation after blastocyst elongation (Fry 1992). LIF 1 and interleukin 11 are necessary for normal development of gestation (Pavia et al. 2009). Likewise, the epidermal and fibroblastic growth factors (Gharib-Hamrouche et al. 1993).

Endogenous retroviruses (ERVs) are present in the vertebrate genome. They originated from exogenous viral infections in the germinative line of the recipient. The evolution of ERVs took place jointly with its host for many centuries, and it is known that they contribute to genome plasticity and protect the host against infections caused by related retroviruses and play a vital role in placental development. The ewe has at least 27 copies of ERVs. There is evidence that the retrovirus enJVRs and the hyaluroglycosanimidase 2 expressed in the ULE and the ovine TR, particularly in the CBN, could intervene in TR morphogenesis and differentiation (Black et al. 2010).

A neurotrophin, the neural factor, is needed for neuronal development and appears to be also important to maintain gestation in mice and humans (Frank et al. 2014).

In the ovine uterus, there are differences about innervation, among histological zones. The nerve fibers are more numerous in the intercaruncular than in the caruncular zones, probably because they are anatomically related to the uterine glands. Furthermore, the number of nerve fibers was found to be higher during diestrus as compared to the estrogenic phase of the estrous cycle. It was attributed to P influence. Nevertheless, during preimplantation and implantation (days 14, 20, and 24) the quantity of fibers gradually diminished (Avella 1997).

Genes MASH-2, HAND-1, IFNT, and PAG-9 are known to promote TR proliferation, differentiation, and function, respectively (Arnold et al. 2006). MASH-2 was related to the generation of TR binucleate cells (Arnold and Fortier 2008).

In summary, a complex and intricate process is brought about during the morphological changes related to implantation and placental development, with an increasing number of molecules identified (Table 10.1).

## *The Chorionic Extremities*

The TR does not attach to the ULE in the extremities of the conceptus. Hence, the chorioallantois floats in the uterine lumen.

Macroscopically, toward the 22th day of gestation in the ewe and the 28th day in the cow, the chorion is seen as detached from the underlining allantois (Hernández 1975).

The TR suffers a process of cell death (Figs. 10.21, 10.22, and 10.23), which advances in the embryo's direction and covers in some cases around 70% of the area occupied by the conceptus. Since allantoic expansion is greater than its vascular development, avascular areas are present in the conceptus before the end of the embryonic period in sheep and cow (Hernández 1975).

This entails chromatin disintegration (Fig. 10.23). As a result, debris components might affect the adjacent endometrium (Fig. 10.22). All the abovementioned changes are compatible with apoptosis. This does not generate inflammatory

**Table 10.1** Some molecules related to trophoblast (TF) adhesion and remodeling, uterine epithelial changes (UE), and uterine and conceptus extracellular matrix modifications during implantation of the conceptus and possible target sites and related proposed processes

| Promoting molecules and site of production | Intermediary molecules | Target structures | Induced process |
|---|---|---|---|
| P ovary | ADAMTS 1 | Matrix EC uterus | Remodeling |
| P ovary | ADAMTS 1 | Allantochorionic matrix | Remodeling |
| Uterine and TF Phosphoprotein and integrins | | TF and UE | Cell membrane adhesion |
| Serines & adhesines | | TF and UE | Cell membrane adhesion |
| Insulin growth factor transporting protein | | TF and UE | Cell membrane adhesion & migration |
| PE interferon tau (ovary and TF) | Ovine endometrial galectins and TF BNC | Conceptus | Chemo-attraction, differentiation, growth, & cell migration |
| Arginine, leucine, & glucose. Uterine glands | | TF | Proliferation and cell migration |
| Lipophosphatidic acid. Endometrium & conceptus | Prostaglandin E2 | TF | Mitosis |
| Uterine epithelia | Vascular adhesion molecule 1 | TF & UE | Adhesion |
| Integrin alfa 4. TF | | TF & EU | Adhesion |
| Bombesin. TR & uterus | | Uterus | Ionic transport and myometrial contraction |
| Retrovirus in JSRVs and hyaluroglycosanimidase 2 | | TF | Formation & differentiation |
| P4 | | FGF 10, insulin growth factors 1 & 3 | Implantation |
| Endothelial growth factor. Uterus & TF | | TF | TF migration |

Information gathered from Bai et al. (2014), Mishra et al. (2013), Froehlich et al. (2012), Kim et al. (2010, 2011), Woclavek-Potocka et al. (2010), Liszewska et al. (2009), Black et al. (2010), Simmons et al. (2009), Satterfield et al. (2006), Pfarrer et al. (2006), Gray et al. (2004), MacIntyre et al. (2002), Johnson et al. (2001), Budipitoj et al. (2001), Beltrán (2000), Matthews et al. (1993), Amiot et al. (1993), and Gharib-Hamrouche et al. (1993)

reactions, as necrosis provokes, but it should be taken into account that during the embryonic period, the conceptus probably does not have a completely developed defense mechanism, as represented by lymphoid and/or myeloid type–derived cells.

In viable conceptuses, the allantois continues growing after the appearance of the cell death process. In the areas where the latter is taking place, the fibroblasts proliferate to form collagen fibers (Fig. 10.22), which are organized to impede the rupture of the extremities of the chorioallantois and its consequent collapse due to allantoic fluid loss.

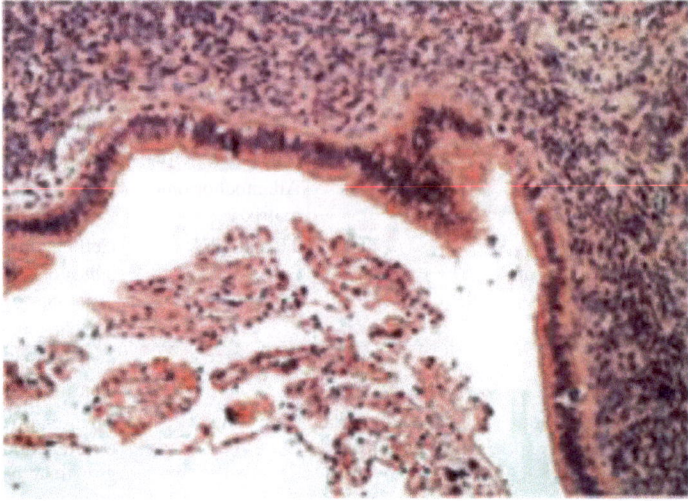

**Fig. 10.21** The bovine chorioallantois is floating in the uterine lumen. The loss of its normal architecture is evident

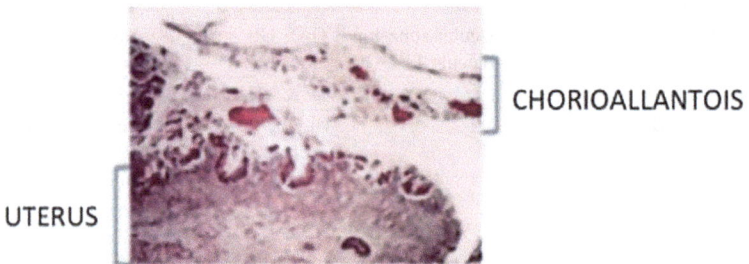

**Fig. 10.22** The necrotic chorioallantois products denude the uterine lining epithelium. (Hernández 1971)

The extent of avascular zones responds to individual variations. It looks like the cell death process covers avascular areas and if vascular supply is not physiologically sufficient to guarantee embryonic survival, conceptus decay might occur (Hernández 1971, 1975). In fact, nonviable conceptuses lack vessels or are avascular (see Fig. 10.33).

Later on in development, collagen is chemically denatured, although not completely destroyed (Hernández 1971; Figs. 10.24 and 10.25).

As gestation advances, the affected areas by cell death are increasingly limited in size and not easily recognizable (Fig. 10.30).

In multiple pregnancies, the chorionic membranes of neighboring conceptuses fuse at the areas previously occupied by the cell death zones. The allantoic cavities of the so fused adjacent conceptuses remain separated (Fig. 10.26), but the allantoic blood vessels may anastomose. The latter would be causing cell and hormonal exchange. Hence, the anti-Müllerian hormone (AMH), dependent on gene

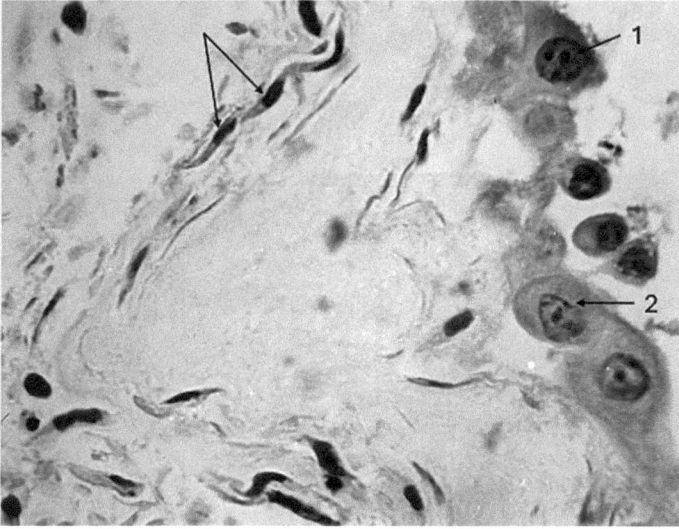

**Fig. 10.23** Apoptotic trophoblast cells. The nuclei have some chromatin remnants (numbers; kary-orrhexis). Fibroblasts (F) are active. (Hernández 1971)

**Fig. 10.24** Photograph of the collagen in the extremities of the conceptus after trophoblast disappearance, as seen with the light microscope. The fibers (F) have not lost their organization, but there is absence of surrounding cells. (Hernández 1971)

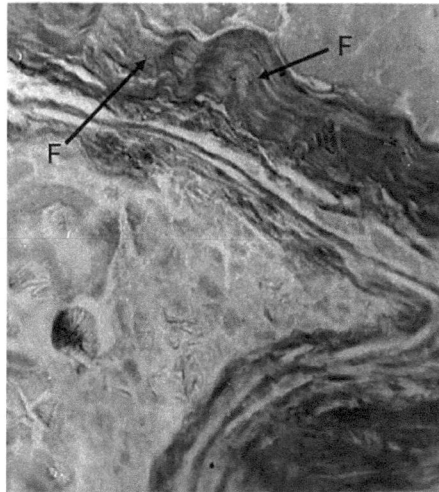

expression within the Y chromosome, which in normal development restrains the paramesonephric (or Müller duct) development in males, would reach the female embryonic system. In cattle cows, it is seen that the development of the parameso-nephric ducts is altered. This intersexuality is known as "freemartin."

The available literature refers to the freemartin condition from diverse points of view, such as cytogenetical, endocrinological, and morphological (see, e.g., Jost et al. (1971) and Laster et al. (1971)).

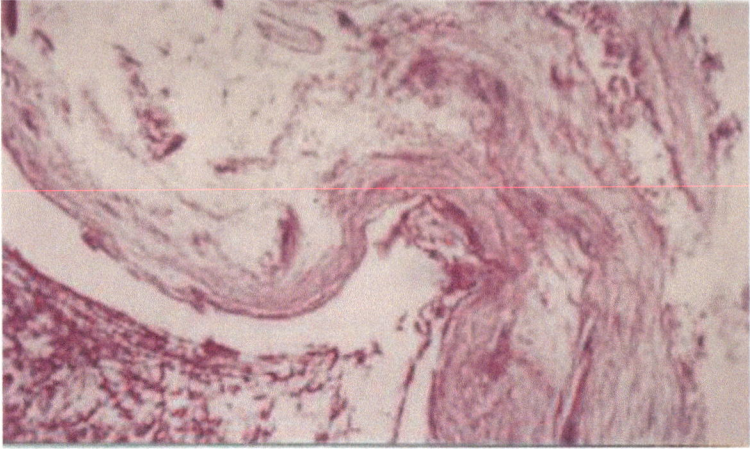

**Fig. 10.25** After trophoblast decay, the collagen in the extremity of the already modified chorioallantois has lost its organization. No cells are seen. PAS stain. (Hernández 1971)

**Fig. 10.26** Induced multiple pregnancy. Sheep. Different contrast media were injected in the allantoic cavities of the conceptuses, as to demonstrate their anatomical independence. Note the variations in size of the conceptuses. (Courtesy of Dr José Rodríguez)

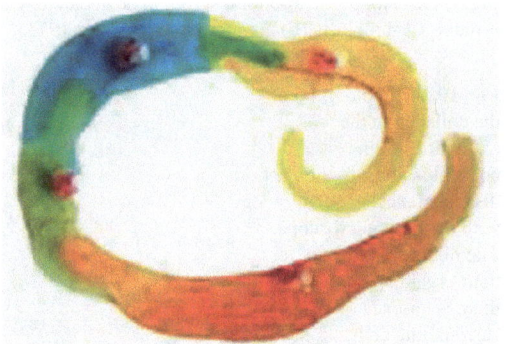

Embryo migration in bovines from the uterine horn adjacent to the ovary containing a corpus luteum (ipsilateral horn) to the opposite uterine horn is not a common finding. Hence, if migration does not occur in double pregnancies, the two conceptuses would probably be nearer to each other, thus increasing the probability of blood vessels anastomosis.

Embryonic migration in sheep is commanded by lysophosphatidic acid, and in pigs, by estrogens (Vallet et al. 2013).

During preimplantation, a series of genes are expressed to guarantee pregnancy success. In the last decade, an important number of studies have been published in the abovementioned topics, but a complete framework is not available at present (see review by Spencer et al. 2008; Rena et al. 2011; Barreto et al. 2011).

Hashizume (2007) reported that 1780 genes were identified as associated with the function of the placenta, specially related with cell organization, cytokines and growth factors elaboration, chaperon and extracellular matrix proteins, enzymes, hormones, and co-enzymes. However, according to the abovementioned author, the function of a third of these molecules is unknown.

The growth of TR through mitosis is a prominent change during its development, but apoptotic and antiapoptotic events are also needed. These are triggered by IFNT during bovine placentation (Groebner et al. 2010). The proliferation of TR cells depends at least in part of induction by growth factor 1, the stimulating colony forming factor, and the endothelial growth factor (Jeong et al. 2014a, b, c).

Some genes of the WNT family codify as to elaborate 19 molecules, which are critical in the differentiation, growth, and cell-to-cell interaction. Some of these genes were identified in the endometrium and TR in sheep, possibly compromised with implantation (Hayashi et al. 2007).

Although the modification of the ULE by the TR in bovines implies less severe changes as compared to what happens in hemo- and endotheliochorial placentae, some common genes were found for the abovementioned types of placentae and the ruminant one. Particularly, genes related to gelatinases—given that their expression augments with the progress of implantation (Hashizume 2007).

The existing epithelial connective tissue in remodeling processes could be exerted through secretion of metalloproteinases (MP). The expression, in the cow, of the extracellular metalloproteinase inducer during embryonic implantation, seems to play a role in TR physiology, particularly in the BNC cells (Mishra et al. 2012). In a study directed to identify MP 2 to 9, and their inhibitor TIMP-2, it was found that M-9 is expressed in the endometrium as well as in the mononucleate cells of the TR, but in the latter together with MP 2. In the BNC, TIMP-2 is expressed, and possibly, it would participate in the detachment of cotyledons from the caruncles near parturition, given that the BNC number diminishes at the end of gestation (Walter and Boos 2001).

The interest in MPs also relays upon their possibly concourse in placental retention (Dilly et al. 2011).

Unbalanced apoptotic mechanisms result in placental defects in humans (Sharp et al. 2010). For instance, SOLD1 was postulated as a controller of fibroblast proliferation and apoptosis in the chorioallantois mesenchyme in the inter- and cotyledonary zones of cows, goats, and sheep (Ushizawa et al. 2010). Hambruch et al. (2010), from their in vitro studies, found that the epidermal growth factor secreted by the cotyledonary TR is needed for its proliferation. Trophoblastic development in the cow is limited by PHLDA gene expressed in the mononuclear TR cells. This expression disappears in individuals obtained by nuclear transfer of somatic cells, beginning at day 200 of gestation, and as a result, placentomegalia takes place (Guillomot et al. 2010). The latter is a failure in placentation seen in individuals obtained by nuclear transfer, due to defects in epigenetic programming (reviewed by Chavatte-Palmer et al. 2012). More genetic abnormalities were detected in conceptuses obtained using in vitro procedures (Talbot et al. 2010; Kohan-Ghadr et al. 2012). Salilew-Wondim et al. (2013) found that 50-day-old fetuses, obtained by nuclear transfer or in vitro fertilization, developed transcription failures of 1196 factors. Nine percent of those individuals showed errors in genetic reprogramming. Genes related to the formation of organ and blood vessels and organization of extracellular matrix and immune system were affected in embryos produced by nuclear transfer and in vitro fertilization. Nevertheless, 96% of altered genes in nuclear transfer obtained embryos was not affected in those individuals obtained by in vitro fertilization.

Arnold et al. (2006) reported disturbed expression of MASH-2 and HAND-1 genes in conceptuses obtained by nuclear transfer or in vitro fertilization.

Palmieri et al. (2008) found a 5% success in cloning processes carried out in humans, mice, and bovines.placentomegaly, absence of blood vessels, and trophoblastic hypoplasia were the most frequent defects reported. During the first trimester of gestation, embryonic losses were 50–70% and 11% if cloning or in vitro fertilization methods were used, respectively. In normal pregnancies, embryonic mortality was found to be 1–3%. For more findings in the same topics, see Arnold and Fortier (2008).

When mammals are cloned, it is expected that the ooplasm eliminates or ignores the methylation patterns established in the nucleus of a differentiated somatic cell. If this process does not take place, anomalies could appear in cloned derived individuals, as already explained (Rideout et al. 2001).

The size of the functional placenta appears to be essential for embryo/fetus nutrition and survival. For example, weight of newborn piglets is correlated with placental size. In cows, embryos prefer the more ample uterine zones for implantation, so that 70% of them are found in the caudal zone of the uterine zones in single pregnancies (Umaña and Hernández 1994; Escobar and Hernández 1996). The same is true in sheep (Gaviria and Hernández 1994).

In ruminants, placentomes are formed by the fusion of uterine caruncles and embryo/fetal cotyledons. In the placentomes, there are more blood vessels, and they are nearer the maternal blood vessels than in the interplacentomary zones. Therefore, in the former, the respiratory function is brought about in a more efficient manner than in the interplacentomary zones, as suggested by Bartels (1970).

In the intercaruncular uterine areas, the glands are essential for embryonic nutrition. Clear-cut evidence was given on this theme from the studies performed in knockout specimens (Spencer and Gray 2006). The growth of the uterine glands during gestation has been well demonstrated (Albers et al. 2015).

Mansouri-Attia et al. (2009) found that there were a difference in gene expression between the caruncular zones and the intercaruncular uterine zones. During gestation, there is greater expression of receptors associated with immune response in the intercaruncular uterine areas than in the caruncular ones (see Oliveira et al. 2012).

Villous formation in the chorioallantois is initiated around day 26th in sheep (Hernández 1971; Fig. 10.27) and on day 28th of gestation in cattle (Aires et al. 2014). Grossly, some cotyledons can be seen by the end of the embryonic period in sheep (i.e., day 32th of gestation) and on day 34th of gestation in the cow (Peter 2013), although as already stated, individual variations must be considered in this context. In fact, Escobar (1993) reported variations in chorioallantois length in five 31-day-old specimens, as follows: 38, 60, 47.3, 34, and 38 cm. Similar findings were published for sheep (Gaviria and Hernández 1994). Henao and Hernández

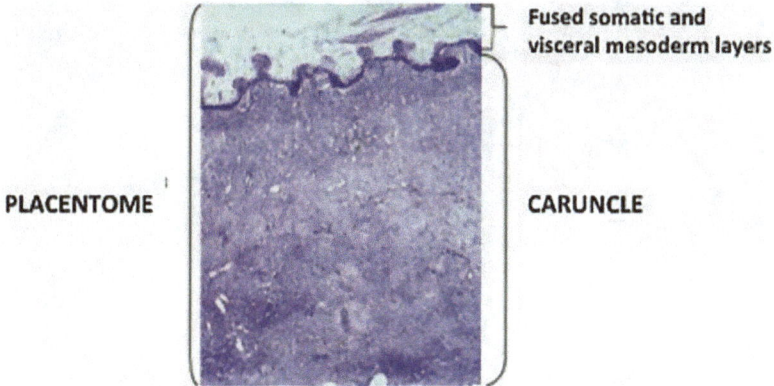

**Fig. 10.27** The sheep chorioallantois, with the already fusioned somatic and visceral mesoderms, projects finger like protrusions in the caruncle. Note the dense vascularity and the absence of glands; 26 days of gestation

(2010) encountered variations in size of pig conceptuses of the same age, of different genotypes.

The first placentomes are seen near the embryo and throughout gestation they are bigger than placentomes occupying more distant areas from the embryo/fetus (Laven and Peters 2001; Figs. 10.29 and 10.30). The placentome number varies; between 37 and 40 days there are as a maximum, 40 placentomes, but during the next 10 days the number is triplicated. After day 70, their number fluctuates between 80 and 90 (Assis Neto et al. 2009). In the placentomes, respiratory gas exchange undergoes and maternal and fetal blood vessels become organized in an intricate and efficient system. Histologically, the diminution in height of the ULE is greater than in the interplacentomary areas, which might account for a shorter diffusion distance of $O_2$ and $CO_2$, as already explained (see Pfarrer et al. 2001).

In the placentomes, the basement membrane of the ULE is in contact with the trophoblastic apical surface (Fig. 10.28), which might give the idea, based on the observation of some placental zones, that the TR erodes all the epithelium. In this context, in the 1950s and mid-1960s, the placentae of sheep and cow were claimed to be of the syndesmochorial type (Amoroso 1952). Assheton (1906) had classified the placenta of the sheep as epitheliochorial. Likewise, in some areas of the ovine placenta where the TR is in contact with the maternal blood (Amoroso 1952; Hernández 1971; Murai and Yamauchi 1986; see Fig. 10.31), it could be thought that the placenta is of the hemochorial type. However, the placenta of ruminants has been classified as epitheliochorial with placentomes, based on in the predominance of TR: ULE relationships in the mature placenta (i.e., at 90 days of pregnancy). In this context, it is currently found that the cow's placenta and sheep placenta are related as "synepitheliochorial," because of the presumptive syncytia formed by binucleate cells and ULE (Wooding 1982). A valid question is to ask how constant and numerous are the syncytia in the placenta throughout pregnancy? Moreover, it

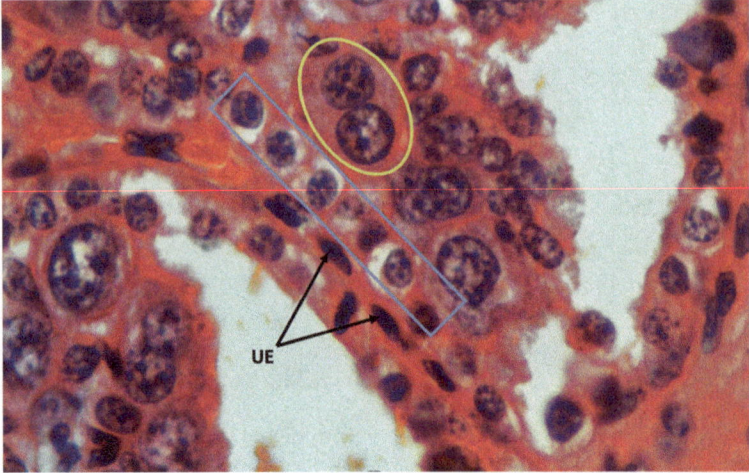

**Fig. 10.28** Placentome, cow. Chorioallantoic villi are cut at different angles. The arrows point to the trophoblast (UE). The oval surrounds a binuclete cell. Rectangular demarcated zone: nuclei of the trophoblast

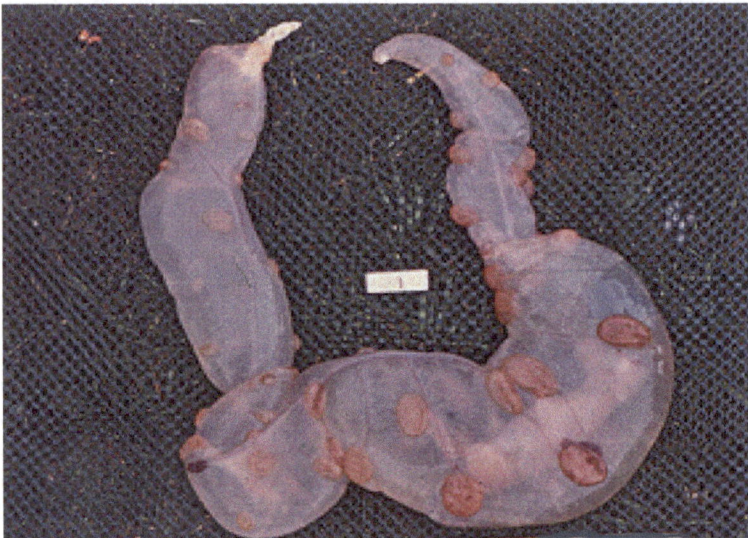

**Fig. 10.29** Bovine conceptus, approximately 80 days of gestation. Cotyledons near the fetus are bigger than those on the periphery. The dead cells zones are limited to yellow scars

is unclear, for how long the binucleate cells remain in the ULE, as to be considered a constant component of the placenta.

It should be noted that the binucleate cells secrete their granules near the basement membrane of the epithelium, and later on, reach the maternal site of the

**Fig. 10.30** Bovine conceptus, 200 days. Note the difference in size of cotyledons. Those near the fetus are bigger than the more distant ones. The dead cells, on the tip zones, are not easily recognizable

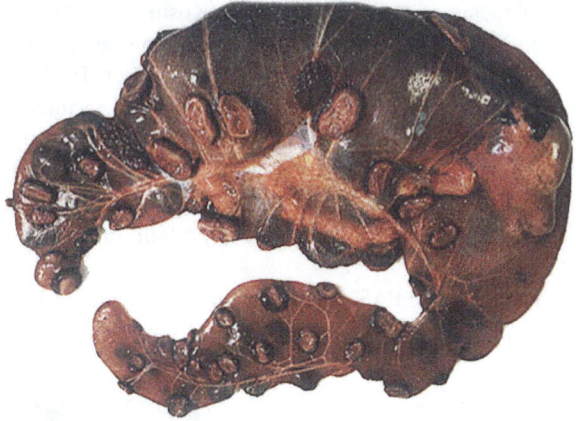

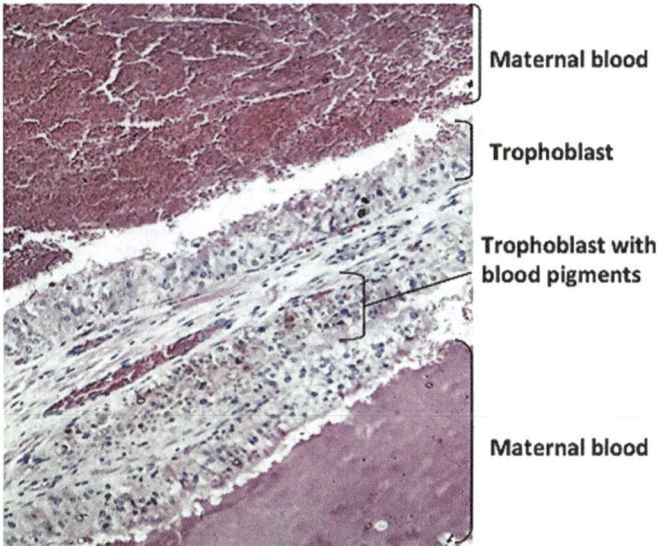

Maternal blood

Trophoblast

Trophoblast with blood pigments

Maternal blood

**Fig. 10.31** Ovine placenta. Eighty days of gestation. The trophoblast is bathed by maternal blood. Although this picture could give the idea of a hemochorial type of placenta, this relationship represents only small zones of the whole placenta. (Hernández 1971)

placenta (Wathes and Wooding 1980; Duello and Gumkowski 1986). Hence, the binucleate cells are just a transitory component of the ULE.

The binucleate cells (BNC) have common characteristics in ruminants (Wooding 1983; Lee et al. 1986; Ushizawa et al. 2007). In sheep, the BNC appearance, around the 16th day of pregnancy, is considered a sign of the initiation of implantation (Boshier 1969). In the cow, the BNC can be seen around the 18th day of gestation and toward the 24th day of gestation the BNC number has augmented (Wathes and Wooding 1980). Possibly, BNC derive from mononucleated cells of the TR through acytokinesis or endoreduplication (Koshi et al. 2012). They represent around 20% of the whole TR cell population throughout gestation (Wooding 1982).

The proliferator activated receptor family (PPAR) and the distal homeobox gene (DLX) are necessary for embryonic survival in mice. They are related to vascular development and placental formation (Barak et al. 1999; Morasso et al. 1999), and they were identified in the cytotrophoblast of the human placenta (Tarrade et al. 2001; Chui et al. 2010). In the bovine TR, evidence was given that both PPAR and DLX act in harmony with SP1 protein in the BNC differentiation (Degrelle et al. 2011).

The BNC are PAS positive (Figs. 10.32 and 10.33) and react with the monoclonal antibody SBU-3 (Lee et al. 1986). The placental lactogen bovine is a

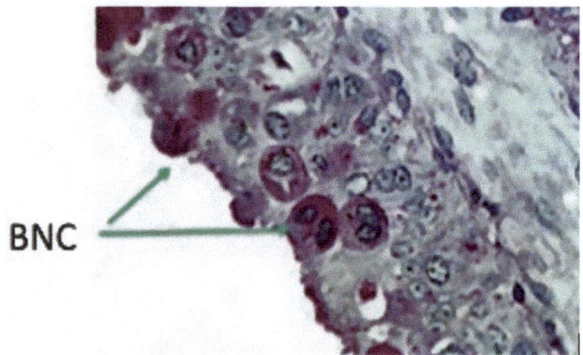

**Fig. 10.32** Binucleate cells in the apical part of the sheep trophoblast. Seventy days of gestation. (Hernández 1971)

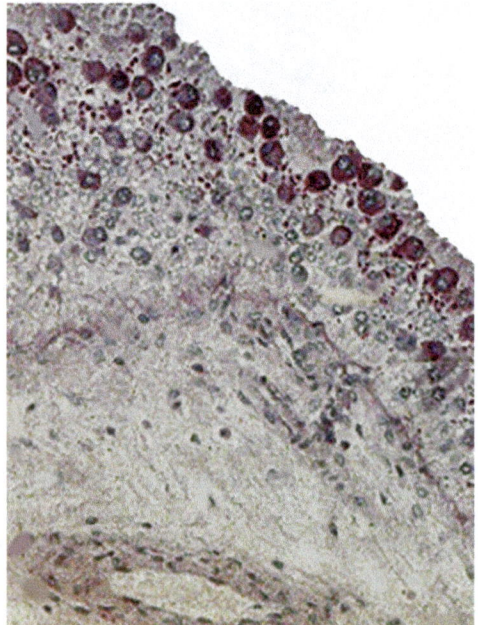

**Fig. 10.33** Sheep trophoblast at the 80th day of pregnancy. Highly stained binucleate cells. PAS and hematoxylin. (Hernández, 1971)

trophoblastic 32K glycoprotein in the BNC, which has three forms that differ in their isoelectric point and the amino acid components. Two of those have similar activity to prolactin and the growth hormone (Duello and Gumkowski 1986).

Duello and Gumkowski (1986) found that the CBN migrate through the ULE. This movement probably implies the fertilin concourse, since it is involved in motility during fertilization, and CD9, which participates in various cell processes, such as proliferation, differentiation, and motility (Xiang and MacLaren 2002).

The BNC placental lactogen can be detected in the maternal blood (Sasser et al. 1986; Zoli et al. 1992; Green et al. 2005). In the mother, it acts in mammogenesis, lactogenesis, and regulation of the ovarian function. It might also act on the fetus (Alvarez-Oxiley et al. 2008).

Another function attributed to the BNC is the control of PG liberation in the mononuclear trophoblastic cells (Gross and Williams 1988). All cells in the TR elaborate estrogens (Matamoros et al. 1994), P, PG I2, and E2 (Reimers et al. 1985; Ullmann and Reimers 1989). PG2 production is likely to be related to a luteotrophic function (Weems et al. 2003). Another claimed function of BNC is the production of adrenomedullin, a vasodilator (Hayashi et al. 2013). They could also participate in extracellular matrix, given their heparanase production (Kizaki et al. 2003).

PGs could facilitate gene expression, which might function in the maternal recognition of pregnancy, from the 13th day pregnancy on (Spencer et al. 2013).

Butler et al. (1982) identified and characterized glycoproteins produced in all the TR cells; they were known as SBU-3 antigen, specific pregnancy associated protein 1, and PSP 60. Nowadays, these ae classified as pregnancy associated glycoproteins (PGA), and two of them, namely, PAG 1 and PGA 9, are identified.

The amount of PAG 9 predominates during the first third of gestation, and PAG 1 does so in later periods (Patel et al. 2004). However, it is currently thought that there are two groups of PGA with variable phylogeny: one of them appeared 87 million years ago and others, 52. The cow's PGA belong to the former. Although PGA belong to the proteinase family, they do not function as enzymes. The most conserved proteins are expressed in the intercotiledonary zones and the least ones in the cotyledons (Touzard et al. 2013).

At least 24 PGA molecules belong to the more recent group. In spite of a fair amount of studies and proposals related to the immune-modulatory or immune-inhibitory and luteotrophic properties of PGA, their function is not clearly defined (reviewed by Wallace et al. 2015).

The lack of proliferator-activated-receptor family and distal homeobox gene (DLX) is lethal in the mouse embryo, due to vascular development defects and placenta formation (Barak et al. 1999; Morasso et al. 1999). DLX and PPAR were identified in cytotrophoblasts of the human placenta (Tarrade et al. 2001; Chui et al. 2010) and bovine TR. In the latter, they would act in conjunction of protein SP1 in the BNC differentiation (Degrelle et al. 2011).

## Some Aspects of Embryonic Nutrition and Metabolism

Studies related to these topics have been brought about using in vitro models related to zygotes or embryos of different species, but particularly in mice.

The Krebs cycle and oxidative phosphorylation are the main energy sources during implantation. Pyruvate is the main substrate used by the embryo in the first stages of development in the majority of animal species. Amino acids, lactate and endogenous fatty acids from triglycerides combined with pyruvate, provide energy during the morula and blastocyst periods. These nutrients have many goals in metabolism. Glucose is mainly used in intracellular outcomes (Leese 2004).

During preimplantation and in some phases of implantation, the TR depends for its survival on the exocrine secretions of the uterine glands (Fig. 10.34). As a whole, their products are known as histotroph, which contains lipids, proteins, and sugars among other compounds (Fig. 10.34). Without the uterine glands, the embryo dies (Gray et al. 2001).

Forde et al. (2014a) identified molecules associated with implantation and maternal recognition of pregnancy and remarked the complexity of the elaboration of uterine fluid substances, related to embryonic nutrition and multiple morphological changes, carried on during the studied stages. Later on, differences in the construction of some amino acid transporting molecules between days 16 and 19 days of gestation were reported (Forde et al. 2014b). Some evidence has been given on the role of amino acids and hexoses in gene transcription and mRNA translation of key molecules in the development of ovine and porcine conceptuses, such as arginine and glucose (Bazer et al. 2012a). In particular, fructose is employed by the embryo/fetus for its development (Bazer et al. 2014). A complete study to identify proteins

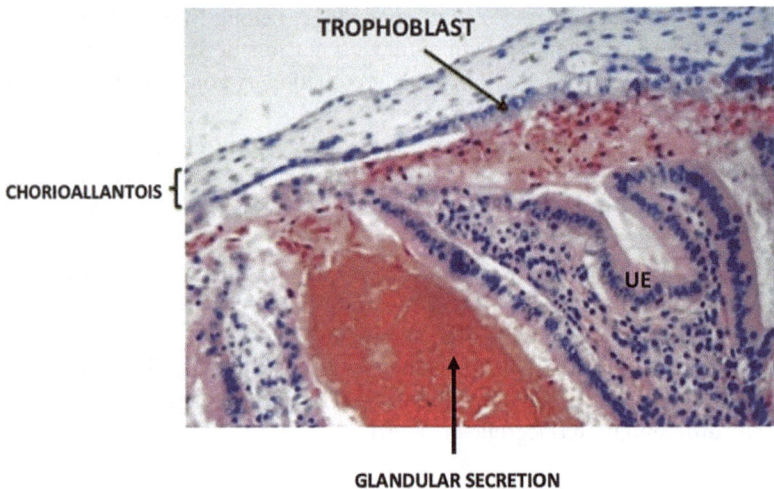

**Fig. 10.34** Placenta of the sheep. Eighty days of gestation. Interplacentomal area. The glandular secretion can be seen in contact with the trophoblast. UE: uterine lining epithelium (unmodified)

in the histotroph during preimplantation was carried out by Mullen et al. (2012). Groebner et al. (2011) provided evidence that the amino acid content in the uterine lumen increases between days 12 and 18 of gestation.

The influence of lactation on embryo survival has been an unsolved question. In a comparative investigation, it was found that lactating steers had lesser live embryos than nonlactating ones. In the latter, blood glucose and insulin growth factor 1 values were lower than in the former group. However, results were not conclusive (Maillo et al. 2012). Similar findings were presented by Green et al. (2012), and in addition, they reported that lactation did not influence serum levels of estrogens and progesterone, but did show a negative effect on fetal and its membranes weight.

Lipogenic diets, especially those with high omega 3 and 6 fatty acids content, represent and aid in embryonic survival (Thatcher et al. 2011). Nevertheless, there are contradictory results (Leroy et al. 2014).

According to Wathes (2012), the mother could derive more nutrients to lactation and less to embryonic support.

It is clear that the embryo and the mother have strategies to favor gestation success even in sub-nutritional conditions, but only to a limit and nutritional failures, negative effects are manifested during gestation and/or extrauterine life (Velazquez 2015).

## Vascular Development in the Allantois

It is generally accepted by clinicians that the embryonic heart beatings begin at 23 days of gestation, although development is not complete as yet. In fact, during the intrauterine life in all organ systems, blood vessels are continuously generated in order to cope with increasing physiological demands.

As already noted, allantoic growth is less rapid than the chorionic one, and therefore, big avascular areas can be seen in conceptuses during implantation, although there is a great deal of variation among individuals. Vascular development in the extraembryonic membranes is not complete before days 60 and 90 in sheep and cows conceptuses, respectively (Hernández 1971; Jiménez and Hernández 1982; Escobar and Hernández 1996. See Fig. 10.35).

De novo construction of blood vessels is known as angiogenesis and that originating from existing ones is called vasculogenesis. A huge amount of molecules intervenes in those processes during placental development. The most studied of them is the vascular endothelial growth factor (VEGF; Eichmann et al. 2005).

The VEGF family includes VEGF A, B, C, D, and E and the placental growth factor. They bind with receptors VEGFR-1 (Flt-1), VEGFR-2 (Flk-1), and VEGFR-3 (Flt-4; Veikkola and Alitalo 1999). The VEGF and its receptor are compromised in the majority of vascular formation processes (Carmeliet et al. 1996; Ferrara et al. 1996).

Vascular permeability is a key event in vascular development. The correspondent receptor in synergy with VEGF acts together during angiogenesis (Murohara et al.

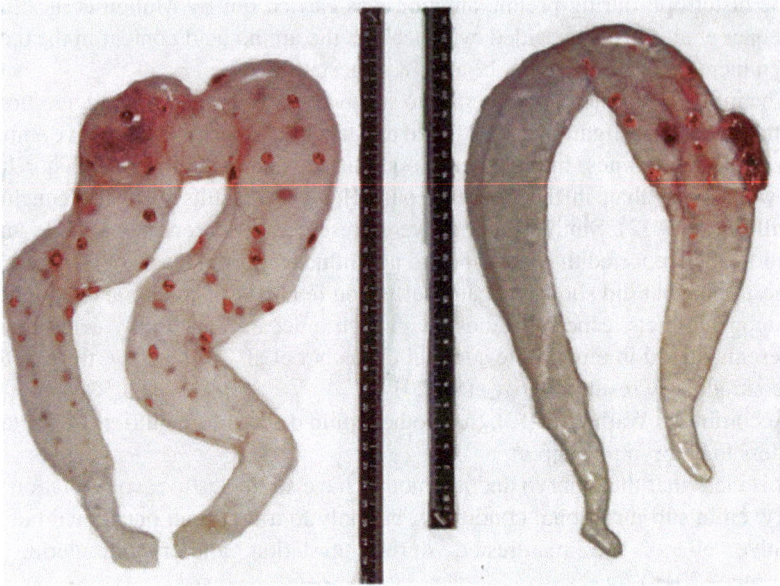

**Fig. 10.35** Two 34-day-old sheep conceptuses. Note the difference in the area occupied by blood vessels. The conceptus on the right side could be of doubtful viability. (Hernández 1971)

1998). VEGF has a role in paracrine maternal: fetal interchange chemotactic activities exerted on the capillary endothelium and during TR migration (Pfarrer et al. 2006).

VEGF expression is higher near the embryo than in the more distant zones at 20, 28, and 35 days of gestation. Similar results were encountered for nitric oxide synthases (endothelial and constitutive). These results were interpreted as a concomitant action of three molecules in vascular development, because VEGF regulates nitric oxide production (Rivas et al. 2007). In advanced stages of gestation, it is feasible that vascular density depends on the expression of other molecules such as cyclooxygenases-2 (Matsumoto et al. 2002), the epidermal growth factor, and the placental growth factor (Athanassiades and Lala 1998). In the human TR, a great amount of angiopogetin-1 was expressed during the second half of gestation (Wulff et al. 2002), which might also occur in ruminants.

The hypoxic environment of the placental environment could stimulate angiogenesis. Angiogenesis is stimulated by the hypoxia-induced factor (HIF) through specific molecule regulation. In particular, HIF-1 directly provokes VEGF and its receptor 1 expression (Liu et al. 1995; Gerber et al. 1997).

Erythroblasts are potentially angiogenic (Tordjman et al. 2001), so that it is likely that during the formation of blood vessels in the primary embryo anatomical entities known as "blood islands" in the allantois, the erythroblasts have a role in angiogenesis and vasculogenesis.

Thirty-day-old conceptuses obtained by cloning or in vitro fertilization are avascular, which is the more salient characteristic of nonviable conceptuses. The avascular condition is possibly due, at least in part, by HIF-1 overproduction, which could induce a diminution in the expression of key molecules for angiogenesis and vasculogenesis (Hoffert-Goeres et al. 2007).

Arginine is a NOS precursor. Wang et al. (2014) found that NOS 3-deficient conceptuses had abnormal development, which might involve a blood vessels formation defect.

Other presumptive participants in the placenta are sphingosine 1 phosphate (Dunlap et al. 2010) and the platelet forming activating factor (Bucher et al. 2006).

Antiangiogenic factors also intervene in blood vessel development regulation in the placenta of the sheep and cow (Reynolds and Redmer 1988; Millaway et al. 1989).

As in ewes and cows, the chorioallantois in pig conceptuses at similar ages has different degrees of vascular development, and the genotype could have an influence (Henao and Hernández 2001; Henao and Hernández 2010). Higher or lower values in the vascular area might imply more or less possibilities of embryonic survival, respectively. However, there is not a reference value related to vascular area as to indicate a critical point in postmortem studies of conceptus viability. Likewise, it has not been established, when the allantoic blood vessels acquire functionality. Nevertheless, nonviable or conceptus of doubtful viability have few blood vessels as a function of their gestational age, or are avascular (Hernández 1971; see Chap. 11).

In the Meishan genotype of pigs, known as a breed of high reproductive performance, it was found that the degree of allantoic vascular density is higher than the correspondent one in commercial breeds. It appears that given that Meishan dams accommodate more embryos than European ones, although the uterine size is similar in all breeds studied, the biological strategy to be more efficient is the higher number of blood vessels per unit area in the allantois (vascular density; Ford 1997).

In the uterus, there should be an important increment in vascular development as gestation progresses to establish a highly efficient respiratory system. In the cow, around 70% of embryos were found located in the caudal part of the uterine horn, ipsilateral to the ovary, which contains the corpus luteum of pregnancy (Umaña and Hernández 1994; Escobar and Hernández 1996). The same is true for single pregnancies in the ewe (Gaviria and Hernández 1994). The abovementioned zones have higher capillary density than other uterine zones (Clavijo and Hernández 1982; Umaña and Hernández 1994).

The distance between the capillary wall in the allantois and the uterine epithelium basal lamina was less around the embryonic zones as compared to more distant ones during implantation. The same was valid in the endometrium of nonpregnant cows. This should imply that the distance of diffusion for respiratory gases diminishes in order to improve efficiency (Umaña and Hernández 1994), since the embryo/fetus is hypoxemic with respect to the mother, and it is feasible that with the advancement of gestation, hypoxemia probably becomes more critical in this context.

During the ovine estrous cycle, it was found that capillary density values were higher in the mid-caudal portion of the uterus, which coincides with the preferential

**Fig. 10.36** Placenta of the sheep. The trophoblast (TR) is in contact with the maternal blood (3). There is evidence of blood components absorption by the trophoblast (arrow). (Hernández 1971)

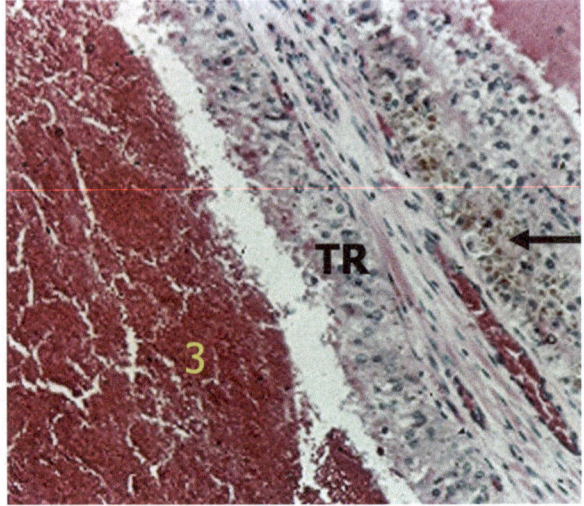

site for embryonic location (Sánchez et al. 2001). It is unknown why the majority embryos locate in that area, but it could be advanced that, in addition of the more abundant vascular irrigation, the caudal zone of the uterine horn covers more area than the cranial ones, due to its greater diameter. Here, a higher number of uterine glands should also be encountered.

In a previous work, it was reported that in the cow, the ipsilateral uterine horn to the ovary which contains the corpus luteum, had more P than the contralateral one (Diaz et al. 1986).

The erythropoiesis during gestation could depend on uteroferrin contained in glandular secretion (Ying et al. 2014). There are restricted ovine placental zones where the TR is in contact with the maternal blood (Fig. 10.36), which may represent a source of Fe for the fetus (Amoroso 1952).

The vascular development and embryo's and placental growth are affected by inadequate supply of nutrients (Redmer et al. 2004).

## Immunological Issues

The uterus needs an immunological depressed environment during gestation, although not completely restricted, because it still has to cope with potential infectious agents (Rosbottom et al. 2011; Oliveira et al. 2012).

During semen passage in the female reproductive tract, or the presence of an embryo, the uterus is conditioned to accept antigenic proteins, as a result of the interaction between cytokines and prostaglandins with the uterine cells' correspondent receptors. Thereafter, genes are expressed as to facilitate spermatozoa viability and avoid immunological rejection of seminal proteins (Schjenken and Robertson 2014) and embryonic antigens.

The conceptus expresses paternal antigens, and embryonic loss can be attributed to the correspondent immunological rejection. For gestation to succeed, the mother should have an efficient immunotolerance. During pregnancy, there is an immuno-logical dialogue between the embryo/fetus and the mother, mediated in part, by the major histocompatibility complex (MHC). The possibilities of conceptus tolerance were attributed to a diminution of paternal type 1 MHC molecules in the TR and to immunosuppression of uterine T lymphocytes activity, exerted by maternal steroids, particularly P (Hansen 2007).

The immunological phenomena become more critical in embryonic transplant-ing cases, due to the double antigenicity exerted either by the male or by the donor female on the recipient female.

The conceptus inhibits the maternal immune response through the expression of IFNT, which, in turn, provokes the elaboration of certain molecules, although some of them are related to disease prevention (Walker et al. 2010).

The mechanisms leading to inactivation of the immune response pertains to the modulation of cell immune response, as related to macrophages and dendritic cells, which control an important number of cytokines (Mansouri-Attia et al. 2012). The intervention of lymphocytes and growth factors is essential. If the mother does not "understand" embryonic messages, there would be immunological rejection. At the end of gestation, the latter occurs as to help in fetal expulsion (Low et al. 1990; Rapacz-Leonard et al. 2014).

To modulate the mother's immunosuppression during gestation and protect her from various infections, a possible mechanism is lymphocyte's inhibition in the uterus, because endometrial lymphocytes are different from others in the organism (Hansen 1995). There is a sufficient number of uterine lymphocytes as to exert immunosuppression (Hansen et al. 1986), and therefore it justifies the need of immunosuppression (Majewski et al. 2001).

The immune competent cells in the cow's endometrium are as follows: CD4+ lymphocytes (helpers), CD8 lymphocytes (cytotoxic), CD5 lymphocytes (B cells), macrophages, and dendritic cells (Cobb and Watson 1995; Miyazawa et al. 2006; Oliveira and Hansen 2009).

The lymphocyte population varies as a function of the age of gestation (Oliveira et al. 2012). In the ULE and the underlying the lamina propria layer, there are lym-phocytes until the 22th day of gestation (King et al. 1981). Vander-Wielen and King (1984) found a drastic reduction in the intraepithelial lymphocytes at the 27th day of pregnancy (Fig. 10.37).

The P immunosuppression mechanism might involve histotroph molecules, without altering the systemic immune response (Hansen 1998). IFNT inhibits lym-phocytes activity in vitro (Newton et al. 1989; Skopets et al. 1992) and the intraepi-thelial lymphocytes are absent by mid-gestation, which is also true for macrophages in the uterine caruncles. A few cells can be seen in the intercaruncular areas (Gogolin-Ewens et al. 1989; Low et al. 1990). From these findings, it can be sug-gested that immunomodulation has a preferential site, in the placentomes. In mice, some lymphocytes appear as necessary for pregnancy success, since their

**Fig. 10.37** Sheep pregnant uterus, 23 days of gestation, intercaruncular zone. In the circle, there some macrophages. The arrows point 2 lymphocytes

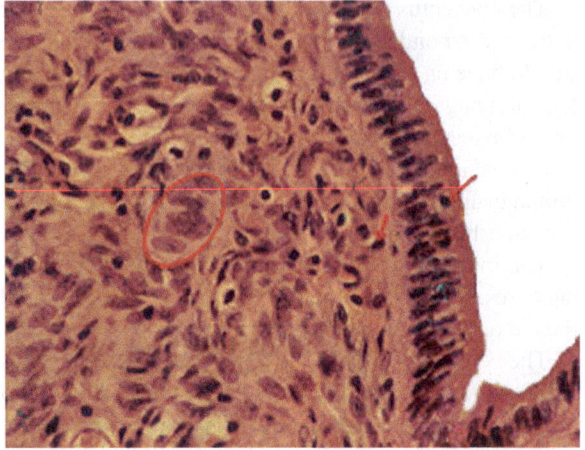

suppression results in miscarriage (Clark et al. 1994). Also, the TR expresses inhibitory genes of cytokine action (Sandra et al. 2005).

In sheep, there is a great deal of macrophages in the endometrial connective tissue during early gestation (Fig. 10.37). It was proposed that these cells could function in tissue remodeling, particularly in fibroblasts' differentiation, angiogenesis, and apoptosis, because they express PDGF beta (Oliveira et al. 2010), although there is not a clear explanation, in this issue, about the presence of the abovementioned cells in the endometrium (Oliveira et al. 2012).

In the sheep uterus, endometrial eosinophils are a normal finding during early pregnancy up to the 28th day of gestation, when they abruptly disappear (Hernández 1971).

In bovines, MHC type 1 proteins are only expressed after the 120th day of gestation (Davies 2007), which poses a question about the chronology of the immune-tolerance mechanism. The BNC might participate, because they produce prostaglandin E2, an inhibitor of T lymphocytes (Low and Hansen 1988).

The expression of the receptor for macrophage inhibitory migration factor augments in the pregnant uterus (Paulesu et al. 2012). Moreover, the lessening of expression of uterine interleukin (IL-2) during gestation could be one of the mechanisms of immune modulation (Leung et al. 2000). This action would be coordinated by IL-1, since this cytokine participates in the cellular immune response during gestation in synergy with the gamma interferon and necrosis tumor factor alpha during implantation. Also, prostaglandin E2 is a possible immune modulator (Wegmann et al. 1993). Apoptosis of immune competent cells induced by trophoblastic cells could also be a mechanism in this context (Abrahams et al. 2004). Also, IFNT inhibits peripheral uterine lymphocytes with a major effect in cows than in steers (Gierek et al. 2006; Ott and Gifford 2010).

In the ewe, a higher number of macrophages were reported in the pregnant than in the nonpregnant uterine horn. Likewise, a relationship was found between immune suppression and serpin in the uterus (Hansen 2007). In the ewe, serpin 14 (initially called milk uterine protein) is the most noticeable histotroph compound. It acts in the immune conceptus acceptance (Padua and Hansen 2010). Its expression is stimulated by estrogen (Ulbrich et al. 2009). Estrogen also inhibits the cell cycle of monocytes and macrophages and P that of promyelocytes in vitro (Thongngram et al. 2003).

During embryo implantation, it is possible to detect in maternal blood cells with immune function acting as IFNT regulators, which might have applications in pregnancy diagnosis (Sandra et al. 2015).

Oliveira et al. (2012) stated that nuclear material from the embryo reaches the maternal circulation. The macrophages could act as intermediary cells to send that material to the maternal circulatory system, which could represent a systemic component of the immune mechanisms during gestation.

In synthesis, the conceptus is not subject of maternal immune rejection, by intermediation of the MHC. Leukocytes and their cytokines also have a role (Davies et al. 2006). In the same way function P4, estrogens and IFNT together with prostaglandin E2, growth factors and IL-1.

# References

Abrahams VM, Straszewski-Chavez SL, Guller S et al (2004) First trimester trophoblast cells secrete Fas ligand which induces immune cell apoptosis. Mol Hum Reprod 10:55–63

Ahn HW, Farmer JL, Bazer FW et al (2009) Progesterone and interferon tau-regulated genes in the ovine uterine endometrium: identification of periostin as a potential mediator of conceptus elongation. Reproduction 138(5):813–825

Aires MB, Dekagi K, Dantzer V et al (2014) Bovine placentome development during early pregnancy. In: Méndez-Vilas A (ed) Microscopy: advances in scientific research and education. Formatex Research Center, Badajoz, pp 390–396

Albers RM, Dekagi K, Dantzer V et al (2015) Quantitative morphological changes in the interplacentomal wall of the gravid uterine horn of cattle during pregnancy. Reprod Biol Endocrinol 18(13):32

Alvarez-Oxiley AV, Sousa NM, Beckers J et al (2008) Native and recombinant bovine placental lactogens. Reprod Biol 8(2):85–106

Amiot F, Leiber D, Marc S et al (1993) GRP-pre-ferring bombesin receptors increase generation of inositol phosphates and tension in rat myometrium. Am J Physiol 265:C1579–C1587

Amoroso EC (1952) Placentation. In: Parkes AS (ed) Marshall's physiology of reproduction, 3rd edn, vol 2. Longmans Green, London

Arnold DR, Fortier AL (2008) Placental insufficiencies in cloned animals – a workshop report. Placenta 29 Suppl A:S108–S110

Arnold DR, Bordignon V, Lefebvre R et al (2006) Somatic cell nuclear transfer alters peri-implantation trophoblast differentiation in bovine embryos. Reproduction 132:279–290

Assheton R (1906) The morphology of ungulate placenta. Particularly the development of that organ in the sheep and notes upon the placenta of the elephant and hyrax. Philos Trans B198:143

Assis Neto AC, Santos ECC, Pereira FTV et al (2009) Initial development of bovine placentation (Bos indicus) from the point of view of the allantois and amnion. Anat Histol Embryol 38:341–347

Athanassiades A, Lala P (1998) Role of placenta growth factor (PlGF) in human extravillous trophoblast proliferation, migration and invasiveness. Placenta 17:545–555

Avella F (1997) Estudio morfofisiológico de la inervación y de las células neuroendocrinas uterinas en el ciclo estral y la implantación temprana en ovinos. Dissertation, Universidad Nacional de Colombia, Bogotá

Bai R, Bai H, Kuse M et al (2014) Involvement of VCAM1 in the bovine conceptus adhesion to the uterine endometrium. Reproduction 148:119–127

Banu SK, Lee J, Stephen SD et al (2010) Interferon tau regulates PGF2 alpha release from the ovine endometrial epithelial cells via activation of novel JAK/EGFR/ERK//EGR-1. Mol Endocrinol 24:2315–2330

Barak Y, Nelson MC, Ong S et al (1999) PPAR gamma is required for placental, cardiac, and adipose tissue development. Mol Cell 4(4):585–595

Barnea ER (2004) Insight into early pregnancy events: the emerging role of the embryo. Am J Reprod Immunol 51:319–322

Barreto RSN, Bressan FF, Oliveira LJ et al (2011) Gene expression in placentation of farm animals: an overview of gene function during development. Theriogenology 76(4):589–597

Bartels H (1970) Chapter 3. In: Prenatal respiration, 1st edn. ED North Holland Publishing Company, Amsterdam

Bazer FW, Spencer TE, Ott TL (1996) Placental interferons. Am J Reprod Immunol 35:297–308

Bazer FW, Kim J, Ka H et al (2012a) Select nutrients in the uterine lumen of sheep and pigs affect conceptus development. J Reprod Dev 58(2):180–188

Bazer FW, Burghardt RC, Johnson GA et al (2012b) Uterine biology in pigs and sheep. J Anim Sci Biotechnol 3(1):23

Bazer FW, Wu G, Johnson GA et al (2014) Environmental factors affecting pregnancy: endocrine disrupters, nutrients and metabolic pathways. Mol Cell Endocrinol 398(1–2):53–68

Beltrán MI (2000) Identificación de células productoras de bombesina en el TR ovino en los días 17, 22 y 24 de la gestación. Dissertation, Universidad Nacional de Colombia, Bogotá

Berg DK, van Leeuwen J, Beaumont S et al (2010) Embryo loss in cattle between days 7 and 16 of pregnancy. Theriogenology 73:250–260

Betteridge KJ (1980) Collection, description and transfer of embryos from cattle 10–16 days after oestrus. J Reprod Fertil 59:205–216

Black SG, Arnaud F, Palmarini M et al (2010) Endogenous retroviruses in trophoblast differentiation and placental development. Am J Reprod Immunol 64:255–264

Boshier DP (1969) A histological and histochemical examination and early placentome formation in sheep. J Reprod Fertil 19:51–61

Brooks K, Burns G, Spencer TE (2014) Conceptus elongation in ruminants: roles of progesterone, prostaglandin, interferon tau and cortisol. J Anim Sci Biotechnol 5(1):53

Bucher K, Juretić D, Geres D et al (2006) Platelet-activating factor receptor (PAF-R) and acetyl-hydrolase (PAF-AH) are co-expressed in immature bovine trophoblast giant cells throughout gestation, but not at parturition. Prostaglandins Other Lipid Mediat 79(1–2):74–83

Budipitoj T, Matsuzaki S, Cruzana MB et al (2001) Immunolocalization of gastrin-releasing peptide in the bovine uterus and placenta. J Vet Med Sci 63(1):11–15

Butler JE, Hamilton WC, Sasser RG et al (1982) Detection and partial characterization of two bovine pregnancy-specific proteins. Biol Reprod 26:925–933

Cammas L, Reinaud PP, Bordas N et al (2006) Developmental regulation of prostacyclin synthase and prostacyclin receptors in the ovine uterus and conceptus during the peri-implantation period. Reproduction 131(5):917–927

Carmeliet P, Ferreira V, Breier G et al (1996) Abnormal blood vessel development and lethality in embryos lacking a single VEGF allele. Nature 380:435–439

Cavanagh AC, Morton H (1994) The purification of early-pregnancy factor to homogeneity from human platelets and identification as Chaperonin 10. Eur J Biochem 222:551–560

Chavatte-Palmer P, Camous S, Jammes H et al (2012) Review: placental perturbations induce the developmental abnormalities often observed in bovine somatic cell nuclear transfer. Placenta 33(Suppl):S99–S104

Chen Y, Green JA, Antoniou E (2006) Effect of interferon-tau administration on endometrium of nonpregnant ewes: a comparison with pregnant ewes. Endocrinology 147:2127–2137

Chui A, Pathiradge NA, Johnson B et al (2010) Homeobox gene distal-less 3 is expressed in proliferating and differentiating cells of the human placenta. Placenta 31(8):691–697

Clark DE, Chaouat G, Mogil R et al (1994) Prevention of spontaneous abortion in DBA/2-mated
CBA/J mice by GM-CSF involves CD8+ T-cell-dependent suppression od natural effector
cells. Cell Immunol 154:143–152

Clavijo E, Hernández A (1982) Diferencias en la vascularización de diferentes zonas del endometrio bovino. Rev Col de Cien Pec 4:39–49

Cobb SP, Watson ED (1995) Immunohistochemical study of immune cells in the bovine endometrium at different stages of the oestrous cycle. Res Vet Sci 59:238–241

Conley AJ, Assis-Neto AC (2008) The ontogeny of fetal adrenal steroidogenesis as a prerequisite
for the initiation of parturition. Exp Clin Endocrinol Diabetes 116:385–392

Cooke FN, Pennington KA, Yang Q et al (2009) Several fibroblast growth factors are expressed
during pre-attachment bovine conceptus development and regulate interferon-t expression
from trophectoderm. Reproduction 137:259–269

Davies CJ (2007) Why is the fetal allograft not rejected? J Anim Sci 85(13 Suppl):E32–E35

Davies CJ, Elridge JA, Fisher PJ et al (2006) Evidence for expression of both classical and non-
classical major histocompatibility complex class I genes in bovine trophoblast cells. Am J
Reprod Immunol 55(3):188–200

Degrelle SA, Murthi P, Evan-brion D et al (2011) Expression and localization of DLX3, PPARG
and SP1 in bovine trophoblast during binucleated cell differentiation. Placenta 32(11):917–920

Demmers KJ, Dereka K, Flint A (2001) Trophoblast interferon and pregnancy. Reproduction
121:41–49

Diaz F, Hernández A, Gil A (1986) Niveles de progesterona y morfometría endometrial durante el
ciclo estral de la vaca. Rev Med Zoot 39:15–22

Dilly M, Hambruch N, Shenabai S et al (2011) Expression of matrix metalloproteinase (MMP)-2,
MMP-14 and tissue inhibitor of matrix metalloproteinase (TIMP)-2 during bovine placentation
and at term with or without placental retention. Theriogenology 75(6):1104–1114

Diskin MG, Morris DG (2008) Embryonic and early foetal losses in cattle and other ruminants.
Reprod Domest Anim 43(suppl 2):260–267

Dlaikan H, Hernández A, Cortes A (1999) Modificación del epitelio de revestimiento del útero y
desarrollo trofoblástico, a los 21, 23, 28 y 36 días de la gestación en la vaca. Arch Med Vet
31(2):197–203

Dorniak P, Bazer FW, Spencer TE (2011) Prostaglandins regulate conceptus elongation and mediate effects of interferon tau on the ovine uterine endometrium. Biol Reprod 84(6):1119–1127

Dorniak P, Bazer FW, Wu G et al (2012) Conceptus-derived prostaglandins regulate endometrial
function in sheep. Biol Reprod 87(1):9, 1–7

Dorniak P, Welsh TH, Bazer FA et al (2013) Cortisol and interferon tau regulation of endometrial
function and conceptus development in female sheep. Endocrinology 154(2):931–941

Duello TM, Gumkowski FC (1986) Immunocytochemical localization of protein antigens in large
sections of tissues embedded in water-soluble embedding media. J Histochem Cytochem
34(4):535–538

Dunlap KA, Kwak H, Burhardt RC et al (2010) The sphingosine 1-phosphate (S1P) signaling
pathway is regulated during pregnancy in sheep. Biol Reprod 82(5):876–887

Dunne LD, Diskin MG, Sreenan JM (2000) Embryo and foetal loss in beef heifers between day 14
of gestation and full term. Anim Reprod Sci 58:39–44

Ealy AD, Yang QE (2009) Control of interferon-tau expression during early pregnancy in ruminants. Am J Reprod Immunol 61:95–106

Ealy AD, Larson SF, Liu L et al (2001) Polymorphic forms of expressed bovine interferon-tau
genes: relative transcript abundance during early placental development, promoter sequences
of genes and biological activity of protein products. Endocrinology 142(7):2906–2915

Eichmann A, Yuan L, Moyon D et al (2005) Vascular development: from precursor cells to
branched arterial and venous networks. Int J Dev Biol 49:259–267

Escobar F (1993) Desarrollo del alantocorion y de la vascularización del bovino (cebú mestizo)
durante la implantación. Dissertation, Universidad Nacional de Colombia, Bogotá

Escobar F, Hernández A (1996) Vascularización, crecimiento alantocoriónico y ubicación del embrión (o feto) durante la implantación en la vaca. Rev Med Vet Zoot 44(1):7–11

Ferrara N, Carver-Moore K, Chen H et al (1996) Heterozygous embryonic lethality induced by targeted inactivation of the VEGF gene. Nature 380:439–442

Ford SP (1997) Embryonic and fetal development in different genotypes in pigs. J Reprod Fertil Suppl 52:165–176

Forde N, Lonergan P (2012) Transcriptomic analysis of the bovine endometrium: what is required to establish uterine receptivity to implantation in cattle? J Reprod Dev 58(2):189–195

Forde SP, Carter F, Fair T et al (2009) Progesterone-regulated changes in endometrial gene expression contribute to advanced conceptus development in cattle. Biol Reprod 81(4):784–794

Forde N, Carter F, Spencer TE et al (2011) Conceptus-induced changes in the endometrial transcriptome: how soon does the cow know she is pregnant? Biol Reprod 85:144–156

Forde N, Mehta JP, Macgettigan PA et al (2013) Alterations in expression of endometrial genes coding for proteins secreted into the uterine lumen during conceptus elongation in cattle. BMC Genomics 14:321

Forde N, McGettingam PA, Mehta JP et al (2014a) Proteomic analysis of uterine fluid during the pre-implantation period of pregnancy in cattle. Reproduction 147(5):575–587

Forde N, Simintras CA, Sturmey R et al (2014b) Amino acids in the uterine luminal fluid reflects the temporal changes in transporter expression in the endometrium and conceptus during early pregnancy in cattle. PLoS One 9(6):e100010

Frank P, Barrientos G, Tirado-Gonzáles I et al (2014) Balanced levels of nerve growth factor are required for normal pregnancy progression. Reproduction 148:179–189

Froehlich R, Hambruch N, Haeger JD et al (2012) Galectin fingerprinting detects differences in expression profiles between bovine endometrium and placentomes as well as early and late gestational stages. Placenta 33(3):195–201

Fry RC (1992) The effect of leukaemia inhibitory factor (LIF) on embryogenesis. Reprod Fertil Dev 4(4):449–458

Gao H, Wu G, Spencer TE et al (2009) Select nutrients in the ovine uterine lumen. ii. Glucose transporters in the uterus and peri-implantation conceptuses. Biol Reprod 80(1):94–104

Garrett JE, Geisert R, Zavy MT (1988) Evidence for maternal regulation of early conceptus growth and development in beef cattle. J Reprod Fertil 84:437–444

Gaviria MT, Hernández A (1994) Morphometry of implantation in the sheep. I. Trophoblast attachment, modification of the uterine lining, conceptus size and embryo location. Theriogenology 41:1139–1149

Gerber H, Condorelli F, Park J et al (1997) Differential transcriptional regulation of the two vascular endothelial growth Factor receptor genes: Flt-1, but not Flk-1, is upregulated by hypoxia. J Biol Chem 272:23659–23667

Gharib-hamrouche N, Chene N, Guillomot M et al (1993) Localization and characterization of EGF/TGF-alpha receptors on periimplantation trophoblast in sheep. J Reprod Fertil 98:385–392

Gierek D, Baczynska D, Ugorsky M et al (2006) Differential effect of IFN-tau on proliferation and distribution of lymphocyte subsets in one-way mixed lymphocyte reaction in cows and heifers. J Reprod Immunol 71(2):126–131

Gilbert SF (2003) Developmental biology. Sunderland, Massachusetts. In: Principios de biología experimental. Biología del desarrollo, Spanish edition. Panamericana, Bogotá, p 70

Gogolin-Ewens KJ, Lee CS, Mercer WR et al (1989) Site-directed differences in the immune response to the fetus. Immunology 66:312–317

Gonella AM (2011) Comparación de los niveles séricos de 17 beta estradiol y progesterona, sus respectivos receptores endometriales y los cambios histológicos del endometrio durante el ciclo estral en vacas lactantes de tres genotipos. Dissertation, Universidad Nacional de Colombia, Bogotá

Gonella AM, Grajales HA, Martinez R et al (2010) Morphometric evaluation of endometrial glands and progesterone serum levels in three different breeds of cattle in the Colombian tropics. In: Proceedings of the 8th international ruminant reproduction symposium, Anchorage

Grajales H, Hernández A (2008) Niveles séricos de progesterona durante los días 0, 5, 10, 15 y 20 del ciclo estral en novillas Simmental x Cebú, Holstein x Cebú, Romosinuano y Cebú, bajo las condiciones del trópico cálido-húmedo Colombiano. Livest Res Rural Dev 20:39

Grajales H, Hernández A, Prieto E (2006) Determinación de parámetros reproductivos basado en los niveles de progesterona en novillas doble propósito en el trópico colombiano. Livest Res Rural Dev 18:144

Gray CA, Taylor KM, Ramsey WS (2001) Endometrial glands are required for preimplantation conceptus elongation and survival. Biol Reprod 64:1608–1613

Gray CA, Adelson DL, Bazer FW et al (2004) Discovery and characterization of an epithelial-specific galectin in the endometrium that forms crystals in the trophectoderm. Proc Natl Acad Sci USA 101(21):7982–7987

Green JA, Parks TE, Avalle MP et al (2005) The establishment of an ELISA for the detection of pregnancy-associated glycoproteins (PAGs) in the serum of pregnant cows and heifers. Theriogenology 63(5):1481–1503

Green JC, Meyer JP, Williams AM et al (2012) Pregnancy development from day 28 to 42 of gestation in postpartum Holstein cows that were either milked (lactating) or not milked (not lactating) after calving. Reproduction 143(5):699–711

Groebner AE, Schulke K, Unterseer U et al (2010) Enhanced proapoptotic gene expression of XAF1, CASP8 and TNFSF10 in the bovine endometrium during early pregnancy is not correlated with augmented apoptosis. Placenta 31(3):168–177

Groebner AE, Rubio-Aliaga I, Schulke K et al (2011) Increase of essential amino acids in the bovine uterine lumen during preimplantation development. Reproduction 141(5):685–695

Gross TS, Williams WF (1988) Bovine placental prostaglandin synthesis: principal cell synthesis as modulated by the binucleate cell. Biol Reprod 38(5):1027–1034

Guillomot M (1995) Cellular interactions during implantation in domestic ruminants. J Reprod Fertil Suppl 49:39–51

Guillomot M, Thagouti G, Constant F et al (2010) Abnormal expression of the imprinted gene Phlda2 in cloned bovine placenta. Placenta 31(6):482–490

Hambruch N, Haeger JD, Dilly M et al (2010) EGF stimulates proliferation in the bovine placental trophoblast cell line F3 via Ras and MAPK. Placenta 31(1):67–74

Hansen PJ (1995) Interactions between the immune system and the ruminant conceptus. J Reprod Fertil Suppl 49:69–82

Hansen PJ (1998) Regulation of uterine immune function by progesterone—lessons from the sheep. J Reprod Immunol 40(1):63–79

Hansen PJ (2007) Regulation of immune cells in the uterus during pregnancy in ruminants. J Anim Sci 85(13 Suppl):E30–E31

Hansen PJ, Bazer FW, Segerson EC Jr (1986) Skin graft survival in the uterine lumen in the uterine lumen of ewes treated with progesterone. Am J Reprod Immunol Microbiol 12:48–54

Hashizume K (2007) Analysis of utero-placental specific molecules and their functions during implantation in the bovine. J Reprod Dev 53:1–11

Hayashi K, Burghardt RC, Bazer FW et al (2007) WNTs in the ovine uterus: potential regulation of periimplantation ovine conceptus development. Endocrinology 148(7):3496–3506

Hayashi KG, Hosoe M, Sakumoto R et al (2013) Temporo-spatial expression of adrenomedullin and its receptors in the bovine placenta. Reprod Biol Endocrinol 11:62

Henao F, Hernández A (2001) Posible relación del Factor de Crecimiento Endotelial Vascular (VEGF) con la mortalidad embrionaria temprana en cerdos. Revista Sistemas de Producción 11(1):46–58

Henao F, Hernández A (2010) Embryonic survival and placental development in the early pregnancy of the pig. Universidad de Caldas, Manizales

Hernández A (1971) The development of the extremities of the placenta of the domestic sheep. Dissertation, University of Bristol, England

Hernández A (1975) Descripción de las extremidades necróticas de la placenta de la vaca. Rev ICA 10(1):235–242

Hernández A, Rodríguez JM (2008) Implantación embrionaria y reconocimiento materno de la gestación. In: Hernández A (ed) Reproducción en la vaca. Fisiología y aplicaciones. Universidad Nacional de Colombia, Bogotá, p 88

Hoffert-Goeres KA, Batchelder CA, Bertolini M et al (2007) Angiogenesis in day-30 bovine pregnancies derived from nuclear transfer. Cloning Stem Cells 9(4):595–607

Hoffmann B, Goes de Pinho T, Schuller G (1997) Determination of free and conjugated oestrogens in peripheral blood plasma, faeces and urine of cattle throughout pregnancy. Exp Clin Endocrinol Diabetes 105:296–303

Humblot P (2001) Use of pregnancy specific proteins and progesterone assays to monitor pregnancy and determine the timing, frequencies and sources of embryonic mortality in ruminants. Theriogenology 56:1417–1433

Imakawa K, Helmer SD, Nephew KP et al (1997) A novel role of GM-CSF: enhancement of pregnancy specific interferon production, ovine trophoblast protein-1. Endocrinology 132:1869–1871

Jeong W, Song G, Bazer FW et al (2014a) Insulin-like growth factor I induces proliferation and migration of porcine trophectoderm cells through multiple cell signaling pathways, including protooncogenic protein kinase 1 and mitogen-activated protein kinase. Mol Cell Endocrinol 384(1–2):175–118

Jeong W, Kim J, Bazer FW et al (2014b) Proliferation-stimulating effect of colony stimulating factor 2 on porcine trophectoderm cells is mediated by activation of phosphatidylinositol 3-kinase and extracellular signal-regulated kinase 1/2 mitogen-activated protein kinase. PLoS One 9(2):e88731

Jeong W, Kim J, Bazer FW et al (2014c) Stimulatory effect of vascular endothelial growth factor on proliferation and migration of porcine trophectoderm cells and their regulation by the phosphatidylinositol-3-kinase-AKT and mitogen-activated protein kinase cell signaling pathways. Biol Reprod 90(3):50

Jiménez L, Hernández A (1982) Morfología del alantocorion bovino entre los 27 y 88 días de gestación. Re Acovez 9:32–44

Johnson MH, Ziomek CA (1981) The foundation of two distinct cell lineages within the mouse morula. Cell 24:71–80

Johnson GA, Bazer FW, Jaeger LA et al (2001) Muc-1, integrin, and osteopontin expression during the implantation cascade in sheep. Biol Reprod 65:820–828

Jost A, Vigier B, Prepin J et al (1971) Freemartins in cattle: the first steps of sexual organogenesis. J Reprod Fertil 29:349–379

Kim JY, Erikson DW, Burghardt RC et al (2010) Secreted phosphoprotein 1 binds integrins to initiate multiple cell signaling pathways, including FRAP1/mTOR, to support attachment and force-generated migration of trophectoderm cells. Matrix Biol 29(5):369–382

Kim JY, Burghardt RC, Wu G et al (2011) Select nutrients in the ovine uterine lumen. VII. Effects of arginine, leucine, glutamine, and glucose on trophectoderm cell signaling, proliferation, and migration. Biol Reprod 84(1):62–69

King GJ, Atkinson BA, Robertson HA (1981) Development of the intercaruncular areas during early gestation and establishment of the bovine placenta. J Reprod Fertil 61(2):469–474

Kizaki K, Yamada D, Nakano H et al (2003) Cloning and localization of heparanase in bovine placenta. Placenta 24(4):424–443

Kohan-Ghadr HR, Smith LC, Arnold DR et al (2012) Aberrant expression of E-cadherin and beta-catenin proteins in placenta of bovine embryos derived from somatic cell nuclear transfer. Reprod Fertil Dev 24(4):588–598

Koshi K, Suzuki Y, Nakaya Y et al (2012) Bovine trophoblastic cell differentiation and binucleation involves enhanced endogenous retrovirus element expression. Reprod Biol Endocrinol 10:41

Laster DB, Turman EJ, Jhonson BH et al (1971) Effect of sex ratio in utero on degree of transformation and chimaerism in neonatal bovine freemartins. J Reprod Fertil 24:361–367

Laven RA, Peters AR (2001) Gross morphometry of the bovine placentome during gestation. Reprod Domest Anim 36(6):289–296

Ledgard AM, Meier S, Peterson AJ (2011) Evaluation of the uterine environment early in pregnancy establishment to characterise cows with a potentially superior ability to support conceptus survival. Reprod Fertil Dev 23(6):737–747

Lee CS, Gogolin-Ewens K, Brandon MR et al (1986) Comparative studies on the distribution of binucleate cells in the placentae of the deer and cow using the monoclonal antibody, SBU-3. J Anat 147:163–179

Leese HJ (2004) Metabolism of the preimplantation embryo: 40 years on. Reproduction 143:417–427

Leiser R (1975) Kontaktaufnahme swischen trophoblast un uterusepithel während der frühen implantation beim rind. Anat Histol Embryol 4:63–86

Leroy JL, Sturmey G, Hoeck VV et al (2014) Dietary fat supplementation and the consequences for oocyte and embryo quality: hype or significant benefit for dairy cow reproduction? Reprod Domest Anim 49(3):353–361

Leung ST, Derecka K, Mann GE et al (2000) Uterine lymphocyte distribution and interleukin expression during early pregnancy in cows. J Reprod Fertil 119:25–33

Liszewska E, Reinaud P, Billon-Denis E et al (2009) Lisophosphatidic acid signaling during embryo development in sheep: involvement in prostaglandin synthesis. Endocrinology 150:422–434

Liu Y, Cox SR, Morita T et al (1995) Hypoxia regulates vascular endothelial growth factor gene expression in endothelial cells. Circ Res 77:638–643

Lonergan P, Fair T, Forde N (2016) Embryo development in dairy cattle. Theriogenology 86(1):270–277

Low BG, Hansen PJ (1988) Immunosuppressive actions of steroids and prostaglandins secreted by the placenta and uterus of the cow and sheep. Am J Reprod Immunol Microbiol 18:71–75

Low BG, Hansen PJ, Drost M et al (1990) Expression of major histocompatibility complex antigens on the bovine placenta. J Reprod Fertil 90:235–243

Luque EM, Torres PJ, De Loredo N et al (2014) Role of ghrelin in fertilization, early embryo development, and implantation periods. Reproduction 148:159–167

MacIntyre DM, Lim HC, Ryan K et al (2002) Implantation-associated changes in bovine uterine expression of integrins and extracellular matrix. Biol Reprod 66:1430–1436

Maddox-Hyttel P, Alexopoulus NI, Vajta G et al (2003) Immunohistochemical and ultrastructural characterization of the initial post-hatching development of bovine embryos. Reproduction 125:607–623

Maillo V, Rizos D, Besenfelder LJ et al (2012) Influence of lactation on metabolic characteristics and embryo development in postpartum Holstein dairy cows. J Dairy Sci 95(7):3865–3876

Majewski AC, Tekin S, Hansen PJ et al (2001) Local versus systemic control of numbers of endometrial T cells during pregnancy in sheep. Immunology 102:317–322

Mann GE, Lamming GE (2001) Relationships between maternal endocrine environment, early embryo development and inhibition of the luteolytic mechanism in cows. Reproduction 121:175–180

Mann GE, Merson P, Fray MD et al (2001) Conception rate following progesterone supplementation after second insemination in dairy cows. Vet J 162:161–162

Mansouri-Attia N, Aubert J, Reinaud P et al (2009) Gene expression profiles of bovine caruncular and intercaruncular endometrium at implantation. Physiol Genomics 39(1):14–27

Mansouri-Attia N, Oliveira LJ, Forde N, et al (2012) Pivotal role for monocytes/macrophages and dendritic cells in maternal immune response to the developing embryo in cattle. Biol Reprod 87(5):2–12

Marrable AW (1971) The embryonic pig: a chronological account. Pitman, New York, pp 14–23

Matamoros RA, Caamano L, Lamb SV et al (1994) Estrogen production by bovine binucleate and mononucleate trophoblastic cells in vitro. Biol Reprod 51(3):486–492

Matthews CJ, Redfern CP, Thomas EJ et al (1993) Bombesin and gastrin-releasing peptide stimulate elec-tronic ion transport in cultured human endometrial epithelial cell layers. Exp Physiol 78:715–718

Matsumoto H, Ma W-G, Daikoku T et al (2002) Cyclooxygenase-2 differentially directs uterine angiogenesis during implantation in mice. J Biol Chem 277:29260–29267

Maurer RR, Echternkamp SE (1982) Hormonal asynchrony and embryonic development. Theriogenology 17:11–22

Michael DD, Alvarez I, Ocón OM et al (2006) Fibroblast growth factor-2 is expressed by the bovine uterus and stimulates interferon-tau production in bovine trophectoderm. Endocrinology 147(7):3571–3579

Millaway DS, Redmer DA, Kirsch JD et al (1989) Angiogenic activity of maternal and fetal placental tissues of ewes throughout gestation. J Reprod Fertil 86(2):689–696

Minhas BS, Ripps BA, Zhu YP et al (1996) Platelet activating factor and conception. Am J Reprod Immunol 35:267–271

Mishra B, Kizaki K, Koshi K et al (2012) Expression of extracellular matrix metalloproteinase inducer (EMMPRIN) and its expected roles in the bovine endometrium during gestation. Domest Anim Endocrinol 42(2):63–73

Mishra B, Koshi K, Kisaki K et al (2013) Expression of ADAMTS1 mRNA in bovine endometrium and placenta during gestation. Domest Anim Endocrinol 45(1):43–48

Miyazawa K, Aso H, Honda M et al (2006) Identification of bovine dendritic cell phenotype from bovine peripheral blood. Res Vet Sci 81:40–45

Morasso MI, Grinberg A, Robinson G et al (1999) Placental failure in mice lacking the homeobox gene Dlx3. Proc Natl Acad Sci USA 96(1):162–167

Mullen MP, Elia G, Hilliard M et al (2012) Proteomic characterization of histotroph during the preimplantation phase of the estrous cycle in cattle. J Proteome Res 11(5):3004–3018

Mullen MP, Bazer FW, Wu G et al (2014) Effects of systemic progesterone during the early luteal phase on the availabilities of amino acids and glucose in the bovine uterine lumen. Reprod Fertil Dev 26(2):282–292

Murai T, Yamauchi S (1986) Erythrophagocytosis by the trophoblast in a bovine placentome. Nihon Juigaku Zasshi 48(1):75–88

Murohara T, Horowitz JR, Silver M et al (1998) Vascular endothelial growth factor/vascular permeability factor enhances vascular permeability via nitric oxide and prostacyclin. Circulation 97(1):99–107

Newton G, Vallet JL, Hansen PJ (1989) Inhibition of lymphocyte proliferation by ovine trophoblast-1 and a high molecular weight glycoprotein produced by the peri-implantation sheep conceptus. Am J Reprod Immunol 19:99–107

Nguyen PT, Coley AJ, Soboleva TK et al (2012) Multilevel regulation of steroid synthesis and metabolism in the bovine placenta. Mol Reprod Dev 79(4):239–254

Noonan FP, Halliday WJ, Morton H et al (1979) Early pregnancy factor is inmunosuppresive. Nature 278:649–651

O'Sullivan CM, Rancourt SL, Liu SY et al (2001) A novel murine tryptase involved in blastocyst hatching and outgrowth. Reproduction 122:61–71

Okumu LA, Forde M, Mamo S et al (2014) Temporal regulation of fibroblast growth factors and their receptors in the endometrium and conceptus during the pre-implantation period of pregnancy in cattle. Reproduction 147:825–834

Oliveira LJ, Hansen PJ (2009) Phenotypic characterization of macrophages in the endometrium of the pregnant cow. Am J Reprod Immunol 62:418–426

Oliveira L J, McClellan S, Hansen P (2010) Differentiation of the endometrial macrophage during pregnancy in the cow. PLoS One 5(10):e13213

Oliveira L, Barreto RSN, Perecin F et al (2012) Modulation of maternal immune system during pregnancy in the cow. Reprod Domest Anim 47:384–393

Ott TL, Gifford CA (2010) Effects of early conceptus signals on circulating immune cells: lessons from domestic ruminants. Am J Reprod Immunol 64:245–254

Ozawa M, Yang QE, Ealy AD (2013) The expression of fibroblast growth factor receptors during early bovine conceptus development and pharmacological analysis of their actions on trophoblast growth in vitro. Reproduction 145(2):191–201

Padua MB, Hansen PJ (2010) Evolution and function of the uterine serpins (Serpina 14). Am J Reprod Immunol 64:265–274

Palmieri C, Loi P, Dellasalda L et al (2008) Review paper: a review of the pathology of abnormal placentae of somatic cell nuclear transfer clone pregnancies in cattle, sheep, and mice. Vet Pathol 45(6):865–880

Patel OV, Yamada O, Kisaki K et al (2004) Quantitative analysis throughout pregnancy of placentomal and interplacentomal expression of pregnancy-associated glycoproteins-1 and -9 in the cow. Mol Reprod Dev 67(3):257–263

Patten BM (1948) Embryology of the pig. McGraw Hill, New York, p p95

Pavia P, Menkhorst E, Salamonsen L, et al (2009) Leukemia inhibitory factor and interleukin-11: Critical regulators in the establishment of pregnancy. Cytokine Growth Factor Rev 20(4):319–328

Paulesu L, Pfarrer C, Romagnoli R et al (2012) Variation in macrophage migration inhibitory factor [MIF] immunoreactivity during bovine gestation. Placenta 33(3):157–163

Pennington KA, Ealy AD (2012) The expression and potential function of bone morphogenetic proteins 2 and 4 in bovine trophectoderm. Reprod Biol Endocrinol 10:12

Perona RM, Wassarman PM (1986) Mouse blastocysts hatch in vitro by using a trypsin-lyke proteinase associated with cells of mural trophoectoderm. Dev Biol 114:42–52

Peter AT (2013) Bovine placenta: a review on morphology, components, and defects from terminology and clinical perspectives. Theriogenology 80(7):693–705

Peters AR (1996) Embryo mortality in the cow. Anim Breed Abstr 64:587–598

Pfarrer C, Ebert B, Miglino MA et al (2001) The three-dimensional feto-maternal vascular interrelationship during early bovine placental development: a scanning electron microscopical study. J Anat 198(5):591–602

Pfarrer CD, Ruziwa SD, Winther H et al (2006) Localization of vascular endothelial growth factor (VEGF) and its receptors VEGFR-1 and VEGFR-2 in bovine placentomes from implantation until term. Placenta 27(8):889–898

Pfeffer PL, Pearton DJ (2012) Trophoblast development. Reproduction 143(3):231–246

Rapacz-Leonard, A, Dąbrowska M, Janowski T et al. (2014). Major histocompatibility complex I mediates immunological tolerance of the trophoblast during pregnancy and may mediate rejection during parturition. Mediators Inflamm 579279

Redmer DA, Wallace JM, Reynolds LP (2004) Effect of nutrient intake during pregnancy on fetal and placental growth and vascular development. Domest Anim Endocrinol 27(3):199–217

Reimers TJ, Ullmann MB, Hansel W (1985) Progesterone and prostanoid production by bovine binucleate trophoblastic cells. Biol Reprod 33(5):1227–1236

Rena V, Flores-Martín J, Angeletti S et al (2011) StarD7 gene expression in trophoblast cells: contribution of SF-1 and Wnt-beta-catenin signaling. Mol Endocrinol 25(8):1364–1375

Reynolds LP, Redmer DA (1988) Secretion of angiogenic activity by placental tissues of cows at several stages of gestation. J Reprod Fertil 83(1):497–502

Rideout WM, Egan K, Jaenisch R et al (2001) Nuclear cloning and epigenetic reprogramming of the genome. Science 293:1093–1098

Rivas PC, Rodríguez-Márquez JM, Hernández A (2007) Número de vasos sanguíneos y expresión de VEGF, iNOS y eNOS en las células mesenquimales de la alantoides en los días 20, 28 y 35 de la gestación de la oveja. Rev Cient, FCV-LUZ 16(4):347–352

Robertson SA (2007) GM-CSF regulation of embryo development and pregnancy. Cytokine Growth Factor Rev 18(3–4):287–298

Robertson SA, Roberts CT, Farr KL et al (1999) Fertility impairment in granulocyte-macrophage colony-stimulating factor-deficient mice. Biol Reprod 60(2):251–261

Robinson RS, Fray MD, Wathes DC et al (2006) In vivo expression of interferon tau mRNA by the embryonic trophoblast and uterine concentrations of interferon tau protein during early pregnancy in the cow. Mol Reprod Dev 73(4):470–474

Rosbottom A, Gibney H, Kaiser P et al (2011) Up regulation of the maternal immune response in the placenta of cattle naturally infected with Neospora caninum. PLoS One 6:e15799

Roussev RG (1996) Embryonic origin of PreImplantation Factor (PIF): biological activity and partial characterization. Mol Hum Reprod 2:883–887

Rueda BR, Naivar KA, George EM et al (1993) Recombinant interferon-t regulates secretion of two bovine endometrial proteins. J Interferon Res 13:303–309

Sadler T (2006) Langman's medical embryology, 10th edn. Lippincot William & Wilkins Inc, USA

Sakurai T, Bai H, Bai R et al (2013a) Down-regulation of interferon tau gene transcription with a transcription factor, EOMES. Mol Reprod Dev 80(5):371–383

Sakurai T, Nakagawa S, Kim M et al (2013b) Transcriptional regulation of two conceptus interferon tau genes expressed in Japanese black cattle during peri-implantation period. PLoS One 8(11):e80427

Salilew-Wondim D, Tesfaye D, Hossain M et al (2013) Aberrant placenta gene expression pattern in bovine pregnancies established after transfer of cloned or in vitro produced embryos. Physiol Genomics 45(1):28–46

Sánchez JA, Rodríguez J, Hernández A (2001) Área capilar sub-epitelial en el endometrio ovino en los días 0 y 14 del ciclo estral y en los 14, 20 y 24 días de gestación. Rev Cient Univ del Zulia Fac de Cienc Vet 11(1):69–74

Sandra O, Bataillon I, Roux P et al (2005) Suppressor of cytokine signalling genes are expressed in the endometrium and regulated by conceptus signals during early pregnancy in the ewe. J Mol Endocrinol 34:637–644

Sandra O, Constant F, Carvalho AV et al (2015) Maternal organism and embryo biosensoring:insights from ruminants. J Reprod Immunol 108:105–113

Sasser RG, Ruder CA, Ivani KA et al (1986) Detection of pregnancy by radioimmunoassay of a novel pregnancy-specific protein in serum of cows and a profile of serum concentrations during gestation. Biol Reprod 35:936–942

Satterfield MC, Hayashi K, Song G et al (2006) Progesterone regulation of preimplantation conceptus growth and galectin 15 (LGALS15) in the ovine uterus. Biol Reprod 75:289–296

Schjenken JE, Robertson SA (2014) Seminal fluid and immune adaptation for pregnancy—comparative biology in mammalian species. Reprod Domest Anim 49 Suppl 3:27–36

Shang W, Doré JE, Godkin JD (1997) Developmental gene expression of procollagen III in bovine extraembryonic membranes during early pregnancy. Mol Reprod Dev 48(1):18–24

Sharp A, Heazel AEP, Crocker IP et al (2010) Placental Apoptosis in health and disease. Am J Reprod Immunol 64(3):159–169

Simmons RM, Ericson D, Kim J et al (2009) Insulin-like growth factor binding protein-1 in the ruminant uterus: potential endometrial marker and regulator of conceptus elongation. Endocrinology 150(9):4295–4305

Skopets B, Li J, Thatcher WW et al (1992) Inhibition of lymphocyte proliferation by bovine trophoblast protein-1 (type 1 trophoblast interferon) and bovine interferon-alpha 1. Vet Immunol Immunopathol 34:81–96

Song G, Bazer FW, Spencer TE (2007) Pregnancy and interferon tau regulate RASD2 and IFIH2 expression in the ovine uterus. Reproduction 133:285–295

Spencer TE, Gray CA (2006) Sheep uterine gland knockout (UGKO) model. Methods Mol Med 121:85–94

Spencer TE, Ott TL, Bazer FW (1996) t-interferon: pregnancy recognition signal in ruminants. Proc Soc Exp Biol Med 213:215–229

Spencer TE, Sandra O, Wolf E et al (2008) Genes involved in conceptus–endometrial interactions in ruminants: insights from reductionism and thoughts on holistic approaches. Reproduction 135:165–179

Spencer TE, Forde N, Dorniak P et al (2013) Conceptus-derived prostaglandins regulate gene expression in the endometrium prior to pregnancy recognition in ruminants. Reproduction 146(4):377–387

Staggs KL, Austin KJ, Johnson GA et al (1998) Complex induction of bovine uterine proteins by interferón-tau. Biol Reprod 59:293–297

Stamatkin CW, Roussev RG, Stout M et al (2011) PreImplantation Factor (PIF) correlates with early mammalian embryo development-bovine and murine models. Reprod Biol Endocrinol 9:63

Talbot NC, Powell M, Caperna J (2010) Proteomic analysis of the major cellular proteins of bovine trophectoderm cell lines derived from IVP, parthenogenetic and nuclear transfer embryos: reduced expression of annexins I and II in nuclear transfer-derived cell lines. Anim Reprod Sci 120(1–4):187–202

Tarrade A, Schoonjans K, Pavan L et al (2001) PPARgamma/RXRalpha heterodimers control human trophoblast invasion. J Clin Endocrinol Metab 10:5017–5024

Texeira MT, Austin KJ, Perry DJ et al (1997) Bovine granulocyte chemotactic protein-2 is secreted by the endometrium in response to interferon tau. Endocrine 6:31–37

Thatcher W, Santos JE, Staples CR (2011) Dietary manipulations to improve embryonic survival in cattle. Theriogenology 76(9):1619–1631

Thongngram T, Jenkins JK, Ndebele K et al (2003) Estrogen and progesterone modulate monocyte cell cycle progression and apoptosis. Am J Reprod Immunol 49:129–138

Tordjman R, Delaire S, Plouët J et al (2001) Erythroblasts are a source of angiogenic factors. Blood 97(7):1968–1974

Touzard E, Reinaud P, Dubois O et al (2013) Specific expression patterns and cell distribution of ancient and modern PAG in bovine placenta during pregnancy. Reproduction 146(4):347–362

Tsumagari S, Kamata J, Takagi K et al (1993) Aromatase activity and oestrogen concentrations in bovine cotyledons and caruncles during gestation and parturition. J Reprod Fertil 98:631–636

Ulbrich SE, Frohlich T, Schulke K et al (2009) Evidence for estrogen-dependent uterine serpin (SERPINA14) expression during estrus in the bovine endometrial glandular epithelium and lumen. Biol Reprod 81(4):795–805

Ullmann MB, Reimers TJ (1989) Progesterone production by binucleate trophoblastic cells of cows. J Reprod Fertil Suppl 37:173–179

Umaña J, Hernández A (1994) Densidad capilar en el útero bovino durante la implantación. Rev Acovez 19:10–12

Ushizawa K, Takahashi T, Hosoe M et al (2007) Gene expression profiles of novel caprine placental lBMC Dev Biol 7:16

Ushizawa K, Takahashi T, Hosoe M et al (2010) Cloning and expression of SOLD1 in ovine and caprine placenta, and their expected roles during the development of placentomes. BMC Dev Biol 10:9

Vallet JL, McNeel AK, Johnson G et al (2013) Triennial reproduction symposium: limitations in uterine and conceptus physiology that leads to fetal losses. J Anim Sci 91(7):3030–3040

Vander-Wielen AL, King GJ (1984) Intraepithelial lymphocytes in the bovine uterus during the oestrous cycle and early gestation. J Reprod Fertil 70:457–462

Veikkola T, Alitalo K (1999) VEGFs, receptors and angiogenesis. Semin Cancer Biol 9:211–220

Velazquez MA (2015) Impact of maternal malnutrition during the periconceptional period on mammalian preimplantation embryo development. Domest Anim Endocrinol 51C:27–45

Veljsted M (2010) Cellular and molecular mechanisms in embryonic development. In: Hyttel P et al (eds) Essentials of domestic animal embryology. Saunders, Edinburgh, p 32

Walker AM, Kimura K, Roberts MR (2009) Expression of bovine interferon-tau variants according to sex and age of conceptuses. Theriogenology 72(1):44–53

Walker CG, Meier S, Littlejohn MD et al (2010) Modulation of the maternal immune system by the pre-implantation embryo. BMC Genomics 11:474

Wallace RM, Pohler KG, Smith MF et al (2015) Placental PAGs: gene origins, expression patterns, and use as markers of pregnancy. Reproduction 149:R115–R126

Walter I, Boos A (2001) Matrix metalloproteinases (MMP-2 and MMP-9) and tissue inhibitor-2 of matrix metalloproteinases (TIMP-2) in the placenta and interplacental uterine wall in normal cows and in cattle with retention of fetal membranes. Placenta 22(5):473–483

Wang X, Frank JW, Xu J et al (2014) Functional role of arginine during the peri-implantation period of pregnancy. II. Consequences of loss of function of nitric oxide synthase NOS3 mRNA in ovine conceptus trophectoderm. Biol Reprod 91(3):59

Wathes DC (2012) Mechanisms linking metabolic status and disease with reproductive outcome in the dairy cow. Reprod Domest Anim 47(Suppl 4):304–312

Wathes DC, Wooding FB (1980) An electron microscopic study of implantation in the cow. Am J Anat 159(3):285–306

Weems YS, Lewis AW, Randel RD et al (2002) Effects of prostaglandins E2 and F2alpha (PGE2; PGF2alpha), trilostane, mifepristone, palmitic acid (PA), indomethacin (INDO), ethamoxytriphetol (MER-25), PGE2 + PA, or PGF2alpha + PA on PGE2, PGF2alpha, and progesterone secretion by bovine corpora lutea of mid-pregnancy in vitro. Chin J Physiol 45(4):163–8

Weems YS, Kim L, Humphreys V et al (2003) Effect of luteinizing hormone (LH), pregnancy specific protein B (PSPB), or arachidonic acid (AA) on ovine endometrium of the estrous cycle or placental secretion of prostaglandins E2 (PGE2) and F2α (PGF2α) and progesterone in vitro. Prostaglandins Other Lipid Mediat 71:55–73

Wegmann TG, Lin H, Guilbert L et al (1993) Bidirectional cytokine interaction in the maternal–fetal relationship: is successful pregnancy a Th2 phenomenon? Immunol Today 14:353–356

Whitley JC, Shulkes A, Salomonsen LA et al (1998) Temporal expression and cellular localization of a gastrin-releasing peptide-related gene in ovine uterus during the estrous cycle and pregnancy. J Endocrinol 157:139–148

Woclavek-Potocka I, Zieba IK, Skarzynski DJ (2010) Lysophospahatidic acid action during early pregnancy in the cow: in vivo and in vitro studies. J Reprod Dev 56:411–420

Wooding FB (1982) The role of the binucleate cell in ruminant placental structure. J Reprod Fertil Suppl 31:31–39

Wooding FB (1983) Frequency and localization of binucleate cells in the placentomes of ruminants. Placenta 4 Spec No:527–539

Wulff C, Wilson H, Dickson SE et al (2002) Hemochorial placentation in the primate: expression of endothelial growth factor, angiopoietins, and their receptors throughout pregnancy. Biol Reprod 66:802–8012

Xiang W, MacLaren LA (2002) Expression of fertilin and CD9 in bovine trophoblast and endometrium during implantation. Biol Reprod 66(6):1790–1796

Yang QE, Giasetti MI, Ealy AD (2011) Fibroblast growth factors activate mitogen-activated protein kinase pathways to promote migration in ovine trophoblast cells. Reproduction 141(5):707–714

Ying W, Wang H, Bazer FW et al (2014) Pregnancy-secreted acid phosphatase, uteroferrin, enhances fetal erythropoiesis. Endocrinology 155(11):4521–4530

Zoli AP, Guibalt LA, Delahaut P et al (1992) Radioimmunoassay of a bovine pregnancy-associated glycoprotein in serum: its application for pregnancy diagnosis. Biol Reprod 46:83–92

# Chapter 11
# Embryonic Survival and Mortality

**Abstract** The current methods used to assess embryonic survival include ultrasonography, quantification of specific molecules in maternal blood as well as determining the progesterone amounts in both milk and serum of the cow.

When considering embryonic mortality, it is important to highlight several significant causes including infectious diseases, high temperatures and humidity, genetic changes, toxic agents, embryo manipulation, cloning, and various biotechnologies. Some graphic examples of non-viable embryos are herein shown.

**Keywords** Placental lactogen · Prolactin related proteins · IFNT · PAG · ISG family genes · Bacteria · Viruses · Toxic plants · Cortisol · Stress · High temperature · Humidity · Cloning · Chromosome errors · Return to estrus

## The Diagnosis of Embryonic Survival

The current methods used to access embryonic development include ultrasonography, which allows to see embryonic heart beating between 25 and 30 days of gestation, and quantification of specific TR molecules in maternal blood. Additionally, techniques such as rectal palpation and the establishment of P amounts in both milk and serum of the cow are used.

Among TR molecules, the pre-implantation factor found in mice, humans and cows could be used to assess embryonic survival (Ramu et al. 2013). Likewise, metaloproteinases (gelatinase and heparinase), placental latogen, prolactin related proteins, IFNT and PAG (Sousa et al. 2006; Hashizume 2007).

---

The methods employed to detect PGA as a tool of pregnancy diagnosis were reported to have an effectiveness ranging from 93% to 95% (Silva et al. 2007; Pohler et al. 2013). However, the amount of PGA in the maternal blood varies, or the correspondent methods do not have sufficient specificity. Nevertheless, they are currently used, employing serum or milk (Wallace et al. 2015). Shahin et al. (2014) used the abovementioned tool in dual-purpose cattle (Simmental and two crossbred types, Uckermark and Aubrac) found that the PGA quantities vary from the 28th day of gestation on. However, the authors considered that the PGA detection procedure is a reliably one. This method was also validated in goats (González et al. 1999).

Identification of mRNA of some ISG family genes, pregnancy success assessment is feasible at the 18th day of gestation (Sandra et al. 2015).

## Embryonic Mortality

It is important to differentiate embryonic mortality (EM) from fertilization or ovulatory failures.

Evaluation of EM in cattle herds can be approached as a presumptive method using the non-return to estrus figure. For the same purpose, ultrasonography and quantification of various hormones and different molecules in the maternal blood can be used.

The embryonic mortality is over 30% in domestic mammals.

In early EM, the ovarian cycle is not extended, whereas in late EM the cycle's duration is extended and the return to estrus happens some days after the normal period of the estrous (ovarian) cycle. It must be kept in mind that normal cycles duration varies according to the individual characteristics in the number of ovarian follicles growth and regression ("waves") before ovulation. It is known that in the cow, ovarian cycles normally include 2 or 3 waves (and sometimes more) and their duration range goes from 18 to 27 days (Sirois and Fortune 1988; Cardozo et al. 1994). In general, in cases of late EM, the cow returns to estrus after the 28th day of pregnancy, which means that the embryo and/or part of the TR was (were) alive in the 16th day of gestation, when the maternal recognition of pregnancy occurs. This statement is made, given the after embryo's demise, parts of the trophoblast may remain functionally attached to the uterine lining epithelium (Hernández, unpublished).

## *Genetic Factors in Embryonic Mortality*

Embryonic alterations during early gestation result from inherited defects, meiotic or fertilization errors. Also, changes in the number and structure of chromosomes may occur which account for abortions (Ayalon 1978; Sha et al. 2003).

The absence of chromosome homology between trans-located chromosomes during synapsis on the meiotic prophase affects chromosome segregation, followed by genetically unbalanced gametes due to duplications or deficiencies with subsequent EM (Gustavsson and Rockborn 1964).

As already mentioned, lack of expression of the granulocyte macrophage stimulating colony forming factor (GM-CSF) in the mother or the embryos is a cause of EM (Robertson et al. 1999).

Genomic imprinting is essential for normal embryonic development. It is understood as the need of expression of two genes, one from the mother and the other from the father, for normal occurrence of certain developmental biological processes to occur. For instance, the gene that codifies for the insulin growth factor II (IGF II) expression should be in the paternal genome and the gene for the protein controlling excess of IGF II should be activated in the mother. If the latter is activated in the father, it will lead to fetal overgrowth and subsequent death (Gilbert 2003).

For more information about different disorders due to genomic imprinting, see Swales and Spears (2005).

There are 28 identified genes which are IFNT induced (or regulated), related to control early gestation events (Mansouri-Attia et al. 2009, 2012: Forde and Lonergan 2012). Deficiency of IFNT production originates EM. IFNT induces the ISG (interferon stimulated gene) 15, a component of the innate immune system in humans, mice, and ruminants (Hansen 2007a). In female mice ISG ($-/-$) subjected to hypoxic conditions, EM rose up (70% increment). The immune response was lower than the correspondent one in ISG ($+/+$) females, taking migration of natural killer cells as a measure (Henkes et al. 2015).

Lack of allantoic blood vessels is a post-mortem indicator of EM (Boshier 1968; Hernández 1971; Rodríguez et al. 2005. Figures 11.1 and 11.2).

As already noted, failure in the expression of genes involved in angiogenesis and/or vasculogenesis should be a potential cause of EM.

In conceptuses devoid of blood vessels, histologically, the trophoblast is atrophic and the mesoderm is undeveloped (Hernández 1971; Fig. 11.3).

Human and mice embryos which were homozygous for the VEGF gene ($-/-$) dye (Vikkula 1996; Torry and Torry 1997; Vuorela et al. 2000). It could also occur

**Fig. 11.1** Non-viable bovine conceptus. Blood vessels are absent. (Courtesy: Dr. Marcelo R del Campo)

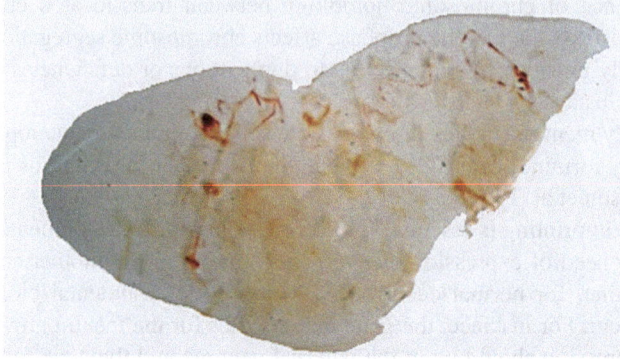

**Fig. 11.2** Cow's chorioallantois. Undetermined age. The embryo died and the chorioallantoic sac was present. A few blood vessels remnants can be seen

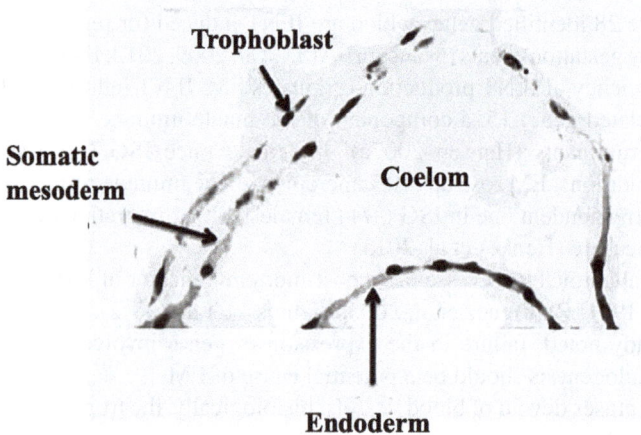

**Fig. 11.3** Histology of the chorioallantois in a dead sheep avascular conceptus. TR: atrophic trophoblast; SM: undeveloped somatic mesoderm. VM: undeveloped visceral mesoderm (Hernández 1971)

in cattle. Likewise, in sheep and pigs since VEGF is expressed in the placenta (Vonnahme et al. 2005; Rivas et al. 2007; Henao and Hernández 2010).

Some genes have direct influence on growth and cell proliferation, lipids metabolism, angiogenesis, and cells and tissue morphology. Failures in their expression result in EM (Killeen et al. 2014).

Failures in normal morphogenesis can cause EM or fetal mortality. For example, Schistosomus reflexus is an inherited defect in ruminants, which results in prenatal death (Laughton et al. 2005). However, many defects are not associated to EM, for instance, Fallot tetralogy (Lacuata et al. 1981; Sinowatz 2010).

# Embryonic Mortality Due to Maladaptation to High Ambient Temperatures

Adaptation to high ambient temperatures is a key issue in reproductive performance. In the northern part of Colombia ("Sinu Valley"), an important meat type cattle industry is located. There the ambient temperature range along the whole year is 20 to 32 Celsius degrees and relative humidity between 72% and 81% (http://bart. ideam.gov.co/cliciu/monter/temperatura.htm). These conditions can be deleterious for reproductive efficiency in non-adapted bovines.

The creole genotype known as "Romosinuano," which has been settled in the Sinu valley for more than 500 years, shows a minor rate of EM as compared to data obtained from non-adapted animals (Hammond et al. 1996). The Romosinuano showed higher conception rate than Zebu, Zebu x Holstein, and Zebu x Simmental cattle (Grajales 2001).

Another creole genotype known as "Sanmartinero" (*Bos Taurus*) showed a better adaptation to high environmental conditions than a *Bos indicus* genotype (Góngora 2001). "Sanmartinero" cattle is settled in the eastern planes of Colombia.

Hernández-Cerón et al. (2004) reported a greater in vitro tolerance to heat in 8-day-old embryos when subjected at 41 °C in Romosinuano and Brahman than Angus and Holstein ones.

Resistance to 41 °C exposure goes up when the zygote passes on to more advanced stages of development (Edwards and Hansen 1997). It was argued that resistance is enhanced with hypoxia, which could imply participation of reactive oxygen molecules or hypoxia-induced compounds (Hansen 2007b).

Cows exposed to 32 °C temperature for 72 hours immediately after artificial insemination had 0% fertility in comparison to cows subjected to temperatures ranging from 7 to 21 °C, which showed 48% fertility (Diskin and Sreenan 1980). Embryos from Brahman cows were more heat resistant than those from Holstein or Angus ones (Paula-Lopes et al. 2003).

There was a 72% IFNT reduction of secretion with the consequent deleterious effect on corpus luteum support, certain prostaglandins secretion, and embryo viability (Putney et al. 1988).

In cattle, the level of adaptation to heat varies according to the genotype and the time of residence in a particular habitat. The association between high readings of relative humidity and high ambient temperatures (temperature:humidity index) originates stress in susceptible individuals and it induces embryonic decay in cows (Santolaria et al. 2010).

Heat stress diminishes oxygen consumption and energy resources and affects mitotic rate in morula stage embryos (Rivera et al. 2004).

## Toxic Agents and Embryo/Fetus Mortality

Phytoestrogens anti-estrogens and some goiter inducers can provoke reproductive disorders (Piotrowska et al. 2006).

Ingestion of *Astragalus* spp. and *Oxytropis* spp. (Conium, Nicotania, Lupinus) plants results in development of fetal malformations, modification in fetal fluids balance, fetal death, and abortion, because they contain pyrimidines (Bunch et al. 1992).

Pinus ponderosa ingestion is a cause of hormonal alterations and/or abortion (Ford et al. 1992).

High consumption of ryegrass (*Lolium perenne*) would lead to EM because of its high protein content. Excess crude protein content in forages affects spermatic and embryonic viability (Gaines 1989).

Leroy et al. (2008) found association between the negative energy balance in high milk production, with EM, caused by low P availability.

High ureic nitrogen also disturbs embryo viability (Rhoads et al. 2006). Ergotamine contained in rye and other cereals threatens normal embryonic development during the first stages (Schuenemann et al. 2005).

Quinolizidine (an alkaloid) is a teratogen in domestic animals, present in common pea, wild pea, and wild blackberry (Keeler 1978; Riet-Correa et al. 2012).

## Infectious Agents and Embryo/Fetus Mortality

The *Leptospira* spp. (Ellis 1986) and *Campylobacter fetus* infections cause embryonic death (Bassan and Ayalon 1971). *Haemophilus somnus* infection may result in fertilization failures and EM (Miller and Barnum 1983; Kiupel and Prehn 1986).

Other reported abortive pathogens, not limited to the embryonic period, include *Nocardia farcinica* (Bawa et al. 2010), *Coxiella burnetti* (Muskens et al. 2012), *Histophilus somni*, *Brucella abortus*, *Arcanobacterium pyogenes*, *Chlamydophila* spp., and *Salmonella* (Givens and Marley 2008).

The negative effects in the embryo's life originated by mastitis bacterial infections are exerted in the hypothalamus, the corpus luteum, or the embryo, through IFN. It causes EM when it is injected between 13 and 19 days of gestation, because it provokes hyperthermia, inhibits luteinizing hormone liberation, and reduces P serum concentrations. Cytokines or their products cause EM in mastitis infections, like interferon gamma and necrosis tumoral factor alpha. The latter and interleukin β stimulate prostaglandin F2α production. Endometrial cell proliferation, oocyte maturation, and embryonic development are negatively affected by the tumoral necrosis factor α, nitric oxide, and prostaglandin F2α (Hansen et al. 2004).

*Mycoplasma* spp. and *Ureoplasma* spp. were associated with reproductive disorders (Givens and Marley 2008). *Ureoplasma diversum* causes infertility in bovines, including EM (Kreplin et al. 1987). In Switzerland, *Paraclamidia* spp. has been reported as a cause of abortion in cows, and its potential as a zoonosis has been remarked, since in human beings it affects the respiratory system (Ruhl et al. 2009; Blumer et al. 2011). This case was also reported in Scotland (Deuchande et al. 2010; Wheelhouse et al. 2012). *Pasterella* spp. is another pathogen involved in abortions (Ward 1990).

Fungal infections may be a cause of abortion (Johnson et al. 1994), particularly, those caused by *Aspergillus fumigatus, Mucor* spp., and *Morteriella wolfii* (Givens and Marley 2008).

Regidor-Cerrillo et al. (2014) induced fetal dead by experimental infection with *Neospora caninum.* The bovines act as intermediary hosts; they eat oocysts which cross the placenta. The final hosts are dogs.

*Trichomona fetus* and *Toxoplasma gondii* affect the embryo (Givens and Marley 2008).

Embryonic viability is highly affected by viral infections; the bovine diarrhea virus (Gard et al. 2007; Hansen et al. 2010) penetrates the zona pellucida and causes embryo degeneration (Apelo and Kanagawa 1989). The bluetongue virus in bovines and ovines cause reproductive disorders (Miller 1991; Rodríguez 1991). Herpesviruses were associated with interruption of gestation (Deim et al. 2006). Pestivirus infection was also reported in this context (McGowan and Kirkland 1995). The Akabane and Aino viruses, generically called Akabane, cross the placenta and cause congenital defects in the fetal nervous system (Murphy 1999).


## *Another Factors*


Any stressful event may lead to embryonic and implantation disturbances.

Unusual changes in the herd's social environment are important. Cortisol is liberated under circumstances such as the presence of dominant females, changes in photoperiod, among several stressors. It acts on the hypothalamus-pituitary-ovary axis.

When the mother is stressed, the blood supply to the sheep embryo/fetus diminishes, which might affect embryonic survival (Rakers et al. 2015).

It has been known for a long time that reduction of P serum levels causes EM (Sreenan and Diskin 1983). The lack of synchrony in hormonal status with the embryonic developmental stage is a cause of EM (Pope 1988). IFNT blood levels act in the same way (Spencer et al. 2008). Robinson et al. (2006) found that 35% of conceptuses cannot produce sufficient amounts of IFNT, which affects embryonic viability. Matsuyama et al. (2012) gave evidence that insufficient IFNT production is related to low embryonic growth and EM.

The ovarian dominant follicle and its microenvironment, the oocyte size, and pre-ovulatory follicle exposure to a LH peak of secretion might influence embryonic survival. All these processes are related with the capacity of the embryo for surviving, oocyte competence to successfully perform meiosis, and adequate P secretion by the corpus luteum (Pohler et al. 2012).

Nutrition and maternal stressors, such as high temperatures, hypoxia, prenatal steroids exposure, and the genotype, influence placental development and other body organs (Reynolds et al. 2010).

In an in vitro study, a positive relationship was reported between degradable protein content of the diet, as found in the rumen, and both urea production and prostaglandin F2a synthesis in uterine cells, which might be harmful for embryonic survival (Butler 1998).

High plasma ureic nitrogen levels are associated with embryonic damage (Rhoads et al. 2006).

Abnormal five-day-old embryos were immersed in uterine fluids with higher than normal glucose, total protein, calcium, magnesium, phosphorus, zinc, and potassium contents (Wiebold 1988). Unbalance in nitric oxide availability and cyclic guanosine monophosphate can lead to apoptosis and embryonic loss (Tranguch et al. 2003). Beltman et al. (2010) found association between lack of expression of prostaglandins, triglycerides, and some immunological related molecules and EM at day 7 of gestation.

As already discussed, many gestations obtained by in vitro fertilization and cloning procedures result in EM, particularly in the latter procedure, due to deficiencies in genetic programming (Guillomot et al. 2010; Talbot et al. 2010; Kohan-Ghadr et al. 2012).

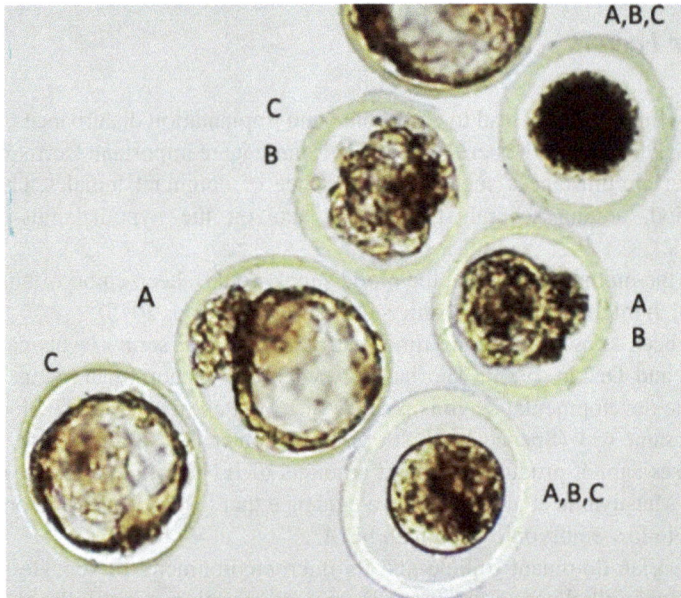

**Fig. 11.4** Non-viable 7-day-old embryos as observed under the light microscope during a transfer process in cows. (**a**) Internal cell mass is displaced. (**b**) Many cell boundaries disappeared. (**c**) The space between the blastocyst and the inner lining of the zona pellucida is bigger than normal

During embryo transplantation ("transfer") procedures, blastocyst evaluation of viability includes morphological issues. Some of them are illustrated in Fig. 11.4.

Rectal palpation before the 60th day of pregnancy can cause EM, due to some present considerations such as conceptus fragility and the slow progression of embryo implantation and placentation.

During evaluation of embryos, many of the individuals obtained by superovulation and subsequent insemination are non-viable (Fig. 11.4).

# References

Apelo CL, Kanagawa H (1989) Pathogens associated with mammalian embryo. Jpn J Vet Res 37(2):49–69

Ayalon NA (1978) A review of embryonic mortality in cattle. J Reprod Fertil 54(2):483–493

Bassan Y, Ayalon N (1971) Abortion in dairy cows inoculated with epizootic bovine abortion agent (Chlamydia). Am J Vet Res 32(5):703–710

Bawa B, Jianfa Bai J, Mike Whitehair M et al (2010) Bovine abortion associated with Nocardia farcinica. J Vet Diagn Invest 22(1):108–111

Beltman ME, Forde N, Furney P et al (2010) Characterisation of endometrial gene expression and metabolic parameters in beef heifers yielding viable or non-viable embryos on Day 7 after insemination. Reprod Fertil Dev 22(6):987–999

Blumer S, Greub S, Waldvogelet A et al (2011) Waddlia, Parachlamydia and Chlamydiaceae in bovine abortion. Vet Microbiol 152(3–4):385–393

Boshier DP (1968) Histological examination of serosal membranes in studies of early embryonic mortality in the ewe. J Reprod Fertil 15:81–86

Bunch TD, Panter KE, James LF (1992) Ultrasound studies of the effects of certain poisonous plants on uterine function and foetal development in livestock. J Anim Sci 70(5):1639–1643

Butler WR (1998) Review: effect of protein nutrition on ovarian and uterine physiology in dairy cattle. J Dairy Sci 81(9):2533–2539

Cardozo J, Díaz F, Hernández A et al (1994) Estrous cycle characteristics and blood progesterone levels in holstein heifers under altitude and tropical conditions in Colombia. Tropicultura 12:148–153

Deim Z, Szeredi L, Tompó V et al (2006) Detection of bovine herpesvirus 4 in aborted bovine placentas. Microb Pathog 41(4–5):144–148

Deuchande R, Gidlow J, Caldow G et al (2010) Parachlamydia involvement in bovine abortions in a beef herd in Scotland. Vet Rec 166(19):598–599

Diskin MG, Sreenan JM (1980) Fertilization and embryonic mortality roles in beef cattle after artificial insemination. J Reprod Fertil 59:463–468

Edwards JL, Hansen PJ (1997) Differential responses of bovine oocytes and preimplantation embryos to heat shock. Mol Reprod Dev 46(2):138–145

Ellis WA (1986) Prevalence of *Leptospira interrogans* serovar *Hardjo* in the genital and urinary tracts of non-pregnant cattle. Vet Res 118(1):11–13

Ford SP, Christenson LK, Rosazza JP (1992) Effects of Ponderosa pine needle ingestion of uterine vascular function in late-gestation beef cows. J Anim Sci 70:1609–1614

Forde N, Lonergan P (2012) Transcriptomic analysis of the bovine endometrium: ¿what is required to establish uterine receptivity to implantation in cattle? J Reprod Dev 58(2):189–195

Gaines JD (1989) The role of record analysis in evaluating subfertile dairy herds. Vet Med 84:532–543

Gard JA, Givens MD, Riddell KP et al (2007) Detection of bovine viral diarrhea virus (BVDV) in single or small groups of preimplantation bovine embryos. Theriogenology 67:1415–1423

Gilbert SF (2003) Developmental biology. Sunderland, Massachisetts. In: Principios de biología experimental. Biología del desarrollo. Séptima edición traducida, (2006). Panamericana. Bogotá, pp 70, 131

Givens MD, Marley MS (2008) Infectious causes of embryonic and fetal mortality. Theriogenology 70(3):270–285

Góngora A (2001) Relación entre factores medioambientales y aspectos reproductivos en novillas Brahman y Sanmartineras del trópico colombiano. Dissertation, Universidad Nacional de Colombia, Bogotá

González F, Sulon J, Garbayo JM et al (1999) Early pregnancy diagnosis in goats by determination of pregnancy-associated glycoprotein concentrations in plasma samples. Theriogenology 52(4):717–725

Grajales HA (2001) Comportamiento reproductivo de grupos raciales bovinos en el trópico cálido-húmedo colombiano: pubertad, ciclo estral, preñez temprana, posparto, niveles de hormonas esteroides y su relación con la eficiencia reproductiva. Dissertation, Universidad Nacional de Colombia, Bogotá

Guillomot M, Thagouti G, Constant F et al (2010) Abnormal expression of the imprinted gene Phlda2 in cloned bovine placenta. Placenta 31(6):482–490

Gustavsson I, Rockborn G (1964) Chromosome abnormality in three cases of lymphatic leukaemia in cattle. Nature 203:990–991

Hammond AC, Olson TA, Chase CC et al (1996) Heat tolerance in two tropically adapted *Bos taurus* breeds, Senepol and Romosinuano, compared with Brahman, Angus, and Hereford cattle in Florida. J Anim Sci 74(2):295–303

Hansen PJ (2007a) Regulation of immune cells in the uterus during pregnancy in ruminants. J Anim Sci 85(13 Suppl):E30–E31

Hansen PJ (2007b) To be or not to be—determinants of embryonic survival following heat shock. Theriogenology 68(Suppl 1):S40–S48

Hansen PJ, Soto P, Natzke RP (2004) Mastitis and fertility in cattle – possible involvement of inflammation or immune activation in embryonic mortality. Am J Reprod Immunol 51(4):294–230

Hansen TR, Smirnova NP, Van Campen H et al (2010) Maternal and fetal response to fetal persistent infection with Bovine Viral Diarrhea Virus. Am J Reprod Immunol 64(4):295–306

Hashizume K (2007) Analysis of utero-placental specific molecules and their functions during implantation in the bovine. J Reprod Dev 53(1):1–11

Henao F, Hernández A (2010) In: Ramirez JF (ed) Embryonic survival and placental development in the early pregnancy of the pig. Universidad de Caldas, Manizales, p 112

Henkes LE, Pru J, Ashley RL et al (2015) Embryo mortality in Isg15−/− mice is exacerbated by environmental stress. Biol Reprod 92(2):36

Hernández A (1971) The development of the extremities of the placenta of the domestic sheep. Dissertation, University of Bristol, England

Hernández-Cerón J, Chase CC Jr, Hansen PJ et al (2004) Differences in heat tolerance between preimplantation embryos from Brahman, Romosinuano, and Angus breeds. J Dairy Sci 87(1):53–58

Johnson CT, Lupson GR, Lawrence KE et al (1994) The bovine placentome in bacterial and mycotic abortions. Vet Rec 134(11):263–266

Keeler RF. In: Kent R, Van Kampen LF, James RF (eds) (1978) Effects of poisonous plants on livestock. Academic Press, New York, pp 397–408

Killeen AP, Morris DG, Kenny DA et al (2014) Global gene expression in endometrium of high and low fertility heifers during the mid-luteal phase of the estrous cycle. BMC Genomics 15:234

Kiupel H, Prehn I (1986) Haemophilus somnus infection of cattle—results of bacteriological study with special reference to abortion substrates. Arch Exp Vet 40(2):164–169

Kohan-Ghadr HR, Smith LC, Arnold DR et al (2012) Aberrant expression of E-cadherin and beta-catenin proteins in placenta of bovine embryos derived from somatic cell nuclear transfer. Reprod Fertil Dev 24(4):588–598

Kreplin CM, Ruhnke HL, Miller RB et al (1987) The effect of intra uterine inoculation with *Ureoplasma diversum* on bovine fertility. Can J Vet Res 51:440–443

Lacuata AQ, Yamada H, Hirose T et al (1981) Tetralogy of Fallot in a heifer. J Am Vet Med Assoc 178:830–836

Laughton KW, Fisher KRS, Halina WG et al (2005) Schistosomus reflexus syndrome: a heritable defect in ruminants. Anat Histol Embryol 34(5):312–318

Leroy JL, Opsomer G, Van Soom A et al (2008) Reduced fertility in high-yielding dairy cows: are the oocyte and embryo in danger? Part I. The importance of negative energy balance and altered corpus luteum function to the reduction of oocyte and embryo quality in high-yielding dairy cows. Reprod Domest Anim 43(5):612–622

Mansouri-Attia N, Aubert J, Reinaud P et al (2009) Gene expression profiles of bovine caruncular and intercaruncular endometrium at implantation. Physiol Genomics 39(1):14–27

Mansouri-Attia N, Oliveira LJ, Forde N et al (2012) Pivotal role for monocytes/macrophages and dendritic cells in maternal immune response to the developing embryo in cattle. Biol Reprod 87(5):2–12

Matsuyama S, Kojima T, Kato TS et al (2012) Relationship between quantity of IFNT estimated by IFN-stimulated gene expression in peripheral blood mononuclear cells and bovine embryonic mortality after AI or ET. Reprod Biol Endocrinol 10(1):1–10

McGowan MR, Kirkland PD (1995) Early reproductive loss due to bovine pestivirus infection. Br Vet J 151(3):263–270

Miller JM (1991) The effects of IBR virus infection on reproductive function of cattle. Vet Med 86(1):95–98

Miller R, Barnum A (1983) Effects of Hemophilus somnus on the pregnant bovine reproductive tract and conceptus following cervical infusion. Vet Pathol 20(5):584–589

Murphy FA (1999) In: Murphy FA, Gibbs EPJ, Horzinek MC, Studdert MJ (eds) Veterinary virology. Academic Press, San Diego, pp 469–483

Muskens J, Wouda W, Bannisseht-Wijsmuller T et al (2012) Prevalence of Coxiella burnetii infections in aborted fetuses and stillborn calves. Vet Rec 170(10):260

Paula-Lopes FF, Chase CC Jr, Al-Katanani YM et al (2003) Genetic divergence in cellular resistance to heat shock in cattle: differences between breeds developed in temperate versus hot climates in responses of preimplantation embryos, reproductive tract tissues and lymphocytes to increased culture temperatures. Reproduction 125(2):285–294

Piotrowska KK, Woclawek-Potocka I, Bah MM et al (2006) Phytoestrogens and their metabolites inhibit the sensitivity of the bovine corpus luteum to luteotropic factors. J Reprod Dev 52(1):33–41

Pohler KG, Geary TW, Atkins JA et al (2012) Follicular determinants of pregnancy establishment and maintenance. Cell Tissue Res 349(3):649–664

Pohler KG, Geary TW, Johnson CL et al (2013) Circulating bovine pregnancy associated glycoproteins are associated with late embryonic/fetal survival but not ovulatory follicle size in suckled beef cows. J Animal Sci 91:4158–4167

Pope WF (1988) Uterine asynchrony: a cause of embryonic loss. Biol Reprod 39(5):99–1003

Putney DJ, Malayer JR, Gross TS et al (1988) Heat stress-induced alterations in the synthesis and secretion of proteins and prostaglandins by cultured bovine conceptus and uterine endometrium. Biol Reprod 39(3):717–728

Rakers F, Bischoff S, Schiffner R et al (2015) Role of catecholamines in maternal-fetal stress transfer in sheep. Am J Obstet Gynecol 213(5):684.e1–684.e9

Ramu S, Stamatkin C, Timms K et al (2013) PreImplantation factor (PIF) detection in maternal circulation in early pregnancy correlates with live birth (bovine model). Reprod Biol Endocrinol 11:105

Regidor-Cerrillo J, Arrans-Solis D, Benavides J et al (2014) Neospora caninum infection during early pregnancy in cattle: how the isolate influences infection dynamics, clinical outcome and peripheral and local immune responses. Vet Res 30(45):10

Reynolds LP, Borowicz PP, Caton JS et al (2010) Developmental programming: the concept, large animal models, and the key role of uteroplacental vascular development. J Anim Sci 88(13 Suppl):E61–E72

Rhoads ML et al (2006) Detrimental effects of high plasma urea nitrogen levels on viability of embryos from lactating dairy cows. Anim Reprod Sci 91(1–2):1–10

Riet-Correa F, Medieros RMT, Schild AL et al (2012) A review of poisonous plants that cause reproductive failure and malformations in the ruminants of Brazil. J Appl Toxicol 32(4):245–254

Rivas PC, Rodríguez-Márquez J, Hernández A (2007) Número de vasos sanguíneos y expresión de VEGF, iNOS y eNOS en las células mesenquimales de la alantoides en los días 20, 28 y 35 de la gestación de la oveja. Revista Científica FCV-LUZ. 16(4):347–352

Rivera RM, Dahigren GM, Augusto de Castro L et al (2004) Actions of thermal stress in two-cell bovine embryos: oxygen metabolism, glutathione and ATP content, and the time-course of development. Reproduction 128(1):33–42

Robertson SA, Roberts CT, Farr KL et al (1999) Fertility impairment in granulocyte-macrophage colony-stimulating factor-deficient mice. Biol Reprod 60(2):251–261

Robinson RS, Fray MD, Wathes DC et al (2006) In vivo expression of interferon tau mRNA by the embryonic trophoblast and uterine concentrations of interferon tau protein during early pregnancy in the cow. Mol Reprod Dev 73(4):470–474

Rodríguez H (1991) Incidencia, Diagnóstico, Control e Impacto Económico de IBR, IPV. Memorias Simposio Leucosis Viral, IBR/VP1. San José de Costa Rica, pp 22–28

Rodríguez J, Hernández A, Hidalgo G (2005) Área vascular del alantocorion ovino: un posible indicador post-mortem de sobrevivencia embrionaria. Rev Cient FCV-LUZ 15(1):14–19

Ruhl S, Casson N, Kaiseret C et al (2009) Evidence for Parachlamydia in bovine abortion. Vet Microbiol 135(1–2):169–174

Sandra O, Constant F. Carvalho AV et al (2015) Maternal organism and embryo biosensoring: insights from ruminants. J Reprod Immunol 108:105–113

Santolaria P, Lopez-Gatius F, García-Ispierto I et al (2010) Effects of cumulative stressful and acute variation episodes of farm climate conditions on late embryo/early fetal loss in high producing dairy cows. Int J Biometeorol 54(1):93–98

Schuenemann GM, Hockett ME, Edwards JL et al (2005) Embryo development and survival in beef cattle administered ergotamine tartrate to simulate fescue toxicosis. Reprod Biol 5(2):137–150

Sha K, Sivapalan G, Gibbons N et al (2003) The genetic basis of infertility. Reproduction 126(1):13–25

Shahin M et al (2014) Pregnancy-associated glycoprotein (PAG) profile of Holstein-Friesian cows as compared to dual-purpose and beef cows. Reprod Domest Anim 49(4):618–620

Silva E, Sterry RA, Kolb D et al (2007) Accuracy of a pregnancy-associated glycoprotein ELISA to determine pregnancy status of lactating dairy cows twenty-seven days after timed artificial insemination. J Dairy Sci 90(10):4612–4622

Sinowatz F (2010) Teratology. In: Ed Hyttel P, Sinowatz F, Veljsted M (eds) Essentials of domestic animal embryology. Saunders, Edinburgh, p 338

Sirois L, Fortune J (1988) Ovarian follicular dynamics in heifers monitored by real-time ultrasonography. Biol Reprod 39:308–317

Sousa NM, Ayad A, Beckers JF et al (2006) Pregnancy-associated glycoproteins (PAG) as pregnancy markers in the ruminants. J Physiol Pharmacol 57(Suppl 8):153–171

Spencer TE, Sandra O, Wolf E et al (2008) Genes involved in conceptus–endometrial interactions in ruminants: insights from reductionism and thoughts on holistic approaches. Reproduction 135(2):165–179

Sreenan JM, Diskin MG (1983) Early embryonic mortality in the cow: its relationship with progesterone concentration. Vet Rec 112(22):517–521

Swales KE, Spears N (2005) Genomic imprinting and reproduction. Reproduction 130(4):389–399

Talbot NC, Powell AM, Caperma TJ et al (2010) Proteomic analysis of the major cellular proteins of bovine trophectoderm cell lines derived from IVP, parthenogenetic and nuclear transfer embryos: reduced expression of annexins I and II in nuclear transfer-derived cell lines. Anim Reprod Sci 120(1–4):187–202

Torry DS, Torry RJ (1997) Angiogenesis and the expression of vascular endothelial growth factor I endometrium and placenta. Am J Reprod Immunol 37(1):21–29

Tranguch S, Steuerwald N, Huet-Hudson YM et al (2003) Nitric oxide synthase production and nitric oxide regulation of preimplantation embryo development. Biol Reprod 68(5):1538–1544

Vikkula M (1996) Dysmorphogenesis caused by an activating mutation in the receptor tyrosine kinase tie. Cell 87:1181–1190

Vonnahme KA, Wilson ME, Li Y et al (2005) Circulating levels of nitric oxide and vascular endothelial growth factor throughout ovine pregnancy. J Physiol 565(1):101–109

Vuorela P, Caroén O, Tulppala M et al (2000) VEGF, its receptors and the tie receptors in recurrent miscarriage. Mol Hum Reprod 6(3):276–282

Wallace RM, Pohler KG, Smith MF et al (2015) Placental PAGs: gene origins, expression patterns, and use as markers of pregnancy. Reproduction 149:R115–R126

Ward AC (1990) Isolation of Pasteurellaceae from bovine abortions. J Vet Diagn Invest 2(1):59–62

Wheelhouse N, Howie F, Gidlow J et al (2012) Involvement of Parachlamydia in bovine abortions in Scotland. Vet J 193(2):586–588

Wiebold JL (1988) Embryonic mortality and the uterine enviroment in first-service lactating dairy cows. J Reprod Fertil 84:393–399

# Index

**A**
Activins, 39, 65, 68
ADN, 24
Aminoacids, 27, 29, 71, 75, 81, 93–95
Angiogenesis, 39, 53, 70, 72, 73, 95–97, 100,
    115, 116
Anti-müllerian hormone (AMH), 12, 21, 24,
    41, 65, 84
Apelin, 40, 53
Aspartate, 29

**B**
Bacteria, 118, 119
Blood supply, 39, 45, 53, 56–57, 59, 61, 119
Bovines, 2–4, 6–8, 14, 28, 65, 66, 71, 75, 76,
    79, 84, 86–88, 90–93, 100,
    115, 117–119

**C**
Cervix, 10–14
Chromosome errors, 114
Cloning, 88, 97, 120
Corpus albicans, 56
Corpus luteum (CL), 36, 37, 51–53, 55, 56,
    71, 72, 74, 86, 97, 117–119
Cortisol, 61, 76, 119
Cow, 11, 36, 37, 45, 46, 60, 63, 65, 66, 70–72,
    77, 79, 81, 82, 87–91, 93, 97–99,
    113, 114, 116

**D**
Differentiation growth factor 9 (GDF9), 39
Dopamine, 31

**E**
Embryo, 1–4, 6, 8–11, 13, 15, 18, 37, 51, 52,
    59, 61, 63–71, 73, 74, 77–79, 82,
    86, 88, 89, 93–99, 101, 114,
    115, 117–120
Embryology, 1–15
Embryo survival, 64, 95
Endorphin β (Eβ), 18, 28, 31, 37
Endothelin (ET-1), 57
Environments, 36, 45, 47, 66, 96, 98, 119
Eosinophils, 45, 100
Estradiol receptors, 28
Estrus, 35, 37, 40, 114

**F**
Fas-L, 57
FHL, 48
Follicle stimulant hormone (FSH), 17, 18, 21,
    23, 24, 27, 28, 31, 36–41, 48, 51, 55
Follicular waves, 35, 37, 51

**G**
Galanin, 28
Gamma interferon, 57, 100